ECOLOGY WARS

Environmentalism As If People Mattered

by

Ron Arnold

A Free Enterprise Battle Book

The Free Enterprise Press

BELLEVUE

Distributed By **MERRIL PRESS**

Acknowledgements

I wish to express my gratitude to the President of the Center for the Defense of Free Enterprise, Alan M. Gottlieb, who helped in many ways; Vance Publishing who released for expansion and updating the original 8-part magazine series "The Environmental Battle" upon which this book is based; to the American Business Press for giving the magazine series their Jesse H. Neal Editorial Achievement Award for 1980.

FIRST EDITION
Published by The Free Enterprise Press

Typeset in Caledonia typeface on Mergenthaler computers by Nova Typesetting Co., Bellevue, Washington.

The Free Enterprise Press is a division of the Center for the Defense of Free Enterprise, 12500 N.E. Tenth Place, Bellevue, Washington 98005.

This book distributed by Merril Press, P.O. Box 1682, Bellevue, Washington 98009. Additional copies of this book may be ordered from Merril Press at $14.95 each.

LIBRARY OF CONGRESS CATALOGING-IN-PUBLICATION DATA

Arnold, Ron.
   Ecology Wars.

   "A Free Enterprise battle book."
   Bibliography: p. 165.
   Includes index.
   1. Environmental policy—United States.
2. Capitalism—United States. I. Title.
HC110.E5A77   1987   363.7'00973   87-17

ISBN 0-939571-00-5

Contents

Dedication:
To
Janet Arnold
my wife, my co-worker and companion
and my children Andrea and Rosalyn
who stuck by me during more thin than thick
And
To those thousands of friends who have helped
along the way and whose help can never be repaid.

Introduction
by Alan M. Gottlieb

For anyone who wants to understand the clash between industry and environmentalists, *Ecology Wars* is the best source you're likely to find. Ron Arnold has lived every word of this dazzling book and he guides us step by step behind the scenes into the heart of environmental conflict.

Ron Arnold was a conservationist years before it became fashionable. He watched his beloved movement shift ground from its traditional "wise use" philosophy into the "grab it all and lock it up" mindset of today's environmental movement. He dared to question the assumptions of his fellow nature-lovers as he saw one environmental protection program after another backfire against the foundations of liberty, private property and free enterprise. When he found answers he didn't like, he did something about it: he challenged the big-government, command-and-control brand of environmentalism as a grassroots free enterprise activist and as an issues manager for American industry.

Ecology Wars is a bomb thrown into the vitals of the entrenched environmental lobby. It reveals things that you've thought for a long time but could never prove. Ron Arnold says there is a better way to protect the environment than to choke America's industrial might to death. Ron Arnold says there are free market ways to control pollution and resource degradation. He says that the environmental movement has betrayed its origins to become another grasping, self-centered "-ism." And he says that environmentalism is no longer about the environment, it's about social change.

If you've ever held back criticism of environmental excesses because you agreed with the goal but not the methods, *Ecology Wars* if for you. Here's the first real explanation of how environmentalists think, of why they do the irrational things they do to protect nature—and why they refuse to consider any solution that might keep free enterprise free.

It's a breath-taking ride through this book: Ron Arnold pulls out all the stops. If you're a dyed-in-the-wool environmentalist and you know how to be honest with yourself, *Ecology Wars* will give you some gut-wrenching moments of self-examination. If you've come to the same conclusions that Ron Arnold has, he'll put a sword in your hand with *Ecology Wars*. No matter who you are, when you come out the other end of this book, you'll never be the same again.

Alan M. Gottlieb, President
Center for the Defense
of Free Enterprise

1

Rethinking Environmentalism

This book is about ecology wars. Ecology wars are fought in legislatures, courtrooms and the hearts and minds of men and women. Keep one fact in mind while you read this book: Ecology isn't environmentalism. Nobody seems to notice, but ecology is a science. Environmentalism is a social movement. An ecologist is a qualified professional in ecology, the science of how organisms interrelate with their environments. An environmentalist is anyone who claims to be one. This is a crucial distinction to keep in mind.

STUDENT TO LOGGER: "What have you got against trees?"

Logger to student: "What have you got against people?"

This exchange overheard in a small northern California cafe symbolizes our ecology wars. When it comes to the environment, emotions flare, discussions of specific problems in specific places stop, and minds grab the closest stereotype for support. It's profit-mad developers against tree-hugging little old ladies in tennis shoes. From the Allagash to Waimea Canyon, personalities and individual belief systems are fully engaged. From the North Slope to the Everglades, the gulf of economic/ecologic misunderstanding is enormous and unbridgeable. In short, ideology, not reason, rules American ecology wars.

And yet everybody is an environmentalist of one kind or another. Long before ecology became a fashionable in-word, environmental awareness glimmered through our very language—think of such timeworn phrases as "don't foul your own nest," for example. But during the 1960s environmental awareness grew to the stature of a mass movement, a cultural change of significant proportions. Sensitivity to nature became first a public virtue, then a requirement, then a fetish. By itself this might have been admirable, or at least amusing or tedious at worst. But America's new-found sensitivity to nature came packaged in a strongly anti-industry, anti-people wrapper. It came with a gut feeling that people are no damn good, that everything we do damages nature and that we must be stopped before we totally destroy the earth. The more radical ecology warriors even thought about shutting down industry for good and all, but the more even-tempered settled for big-government regulation of private industry—and they won in Congress and the courts.

Of course, the private sector didn't like regulation much and fought the "command and control" approach to environmental protection tooth and nail. The more enlightened captains of industry felt that there had to be a way to protect nature without damaging free enterprise, that there were market-oriented solutions to pollution, and that people could live in harmony with nature even in the midst of high-growth industrial civilization. Their ideas were not heard. Organized environmentalists didn't want the public distracted from their big-government regulatory proposals and the media did not sympathize with corporate capitalists. The best that some corporate capitalists could do was try to co-opt the environmental movement with massive foundation grants that in fact only sold rope to the hangmen. And so we got ecology wars. We still have them.

For the past fifteen years I've struggled to understand these wars between environmentalists and industry, perhaps harder than most. For up to the time in 1968 when Redwood National Park was carved from private property—the first time such a thing happened in United States history—I had been a staunch proponent of the conservation movement. But ten years later when the Redwood National Park *Expansion* bill took *more* private property from supposedly free enterprises, I found myself firmly advocating the industry viewpoint. To my surprise, I realized that in the ten-year interval *I* had not changed, the *movement* had.

ENVIRONMENTALISM'S "NEW LOOK"

The original conservation movement with its message of wise use had evolved into the environmental movement with its activism for endless regulation. In the tumultuous and sometimes bitter journey between these two polarized outlooks I encountered many environmental activists, politicians, and industry managers. I have seen both sides of many issues, been torn between loyalties, and made many hard decisions. Most importantly, as a writer I turned for information to natural scientists who dispassionately studied ecology and to social scientists who dispassionately studied mass movements. I can now lay claim to some understanding of the conflict.

The time has come to explode some of the environmental myths that I myself helped create, because they have gone too far: they now constitute a clear and present danger to the survival and well-being of our national economy and our open society. In the past fifteen years, more than a hundred environmental regulatory laws have been passed that focus primarily on social rather than economic results—and the public is not aware of either the laws themselves or of the tremendous costs they levy in higher prices, reduced industrial vigor and outright business failures. Even in the middle of the Reagan administration's first term, in 1982, regulations cost the American economy over $125 billion. Of this amount, 83 percent was spent on compliance with "social engineering" objectives, on bureaucrats playing doctor with natural resources and human lives.

That same year, 1982, even under a conservative Republican president, the United States imported 28 percent of its oil at a cost of $62 billion—a contributing factor to U.S. trade deficits that have made us a debtor nation for the first time in half a century. Ironically, experts estimate that 50 percent of all America's known energy reserves, coal, oil and natural gas, lie under "public" lands, government property that is being methodically sealed off from economic use by overzealous preservationists in the name of "wilderness" and "national parks"—and again, the public is not aware of what's going on. How many people, for example, realize that America will

11

never achieve energy independence as long as the environmental lobby holds sway over Congress?

By 1985, the environmental lobby—a lobby, incidentally, with some of the biggest "clout" in Washington—had pressured Congress to lock up more than 80 million acres of America in federal wilderness areas, with an additional 77 million acres in federal parks. Since the environmental movement began, most of these "preserved" areas were formed not as traditional national parks and reserves were, from existing federal lands, but were seized by government force from private property owners by Act of Congress, a silent scandal that threatens the very roots of American property rights and civil rights.

Even outdoor recreation has been damaged by over-restrictive environmental laws. Dozens of impact-tolerant areas have been placed off-limits to motorized vehicles, 4-wheel-drive clubs have been banned from public lands, trail bikes and snowmobiles prohibited from perfectly acceptable places, float planes from long-used lakes, motor boats from federalized rivers, and the motoring public totally denied access to any federal wilderness area whatsoever—all because of the lobbying power of the environmental movement.

THE IMPACT OF CONFLICT

Let's examine the impact of ecology wars on a specific industry: forest products. Of the 3.6 million square miles that make up the United States, 32 percent, or nearly one-third, are covered with forests—a grand total of 1.13 million square miles of trees, living organisms perfectly adapted to capture solar energy by photosynthesis and store it in the form of wood. This huge genetic reservoir contains 865 species, 61 varieties and 101 hybrids (yet only 120 species have any commercial value) that combine into about 146 forest types in 6 major regions. About one-third of the American forest has no commercial value.

Here's a profile of the two-thirds of our forests that *do* have commercial value: Twenty-eight percent of all commercial forests are publicly owned—17 percent in national forests and 9 percent in state and local forests. Seventy-two percent of the commercial timberland in the U.S. is privately owned, with the bulk held not by big business, but by small individual woodlot owners. Only 14 percent of all private forests (3 percent of the total land area of the U.S.) is owned by the forest industry. Yet this small 14 percent of the working forest produces 28 percent of all the wood grown in America, because of free enterprise and market-oriented management practices.

Lest we fail to see the forest for the trees, we should realize that *land*, not trees, is *the* basic forest resource, which we shall emphasize in a few paragraphs. As foresters are fond of pointing out, trees may come and go, but the land remains to grow new forests. In other words, land is the basic means of production of forests. Since there is only so much land in America that will grow trees, which cannot be significantly enlarged, and since there are more users and uses for the land than there is land to go around, conflicts and competition are inevitable. Therefore, all specific forest environmental disputes such as those over clearcutting, use of herbicides, wilderness preservation, protection of endangered species and so on, can be best understood as **disputes between people who want to use forests for different purposes.**

Two significant problems have been created for the public at large by this Wilderness blitz: First, the majority of the recreational public that has thrown its support behind wilderness proposals did so without being aware of the restrictions that this designation entails. They actually wanted recreation areas that allowed various degrees of development from pure Wilderness to backcountry with primitive hiking shelters to "frontcountry" with hostels containing amenities for hikers to regularly developed car camping sites. According to surveys, the public really doesn't mind timber harvest when properly managed along with recreational values, but they're getting neither the kind of recreation areas they want nor the benefits of the timber that could have been converted into useful products on the same lands.

Second, when an area is designated as official Wilderness, it acts like a mating call to the true wilderness lover, and hordes of people rush to share solitude together. This both ruins the wilderness experience (I know, I've suffered from this "Intrusion Factor"), and puts an unbearable load on the

13

area's natural carrying capacity, trampling endemic wild-flowers and frightening away endangered species with armies of nature lovers. The truth is, the American wilderness is being loved to death by its friends. As Colin Fletcher, high priest of backpacking, said in 1971, "The woods are overrun and sons of bitches like me are half the problem." The other half of the problem is the hiker who won't admit being a problem.

Some "wildernists" (a social scientist's neologism for "wil-derness-purist," not name-calling on my part) think they can solve the problem by setting aside ever-increasing amounts of wilderness, and capitalize on the scarcity angle in their prop-aganda. More thoughtful conservationists realize that some kind of regulation on wilderness entry is essential to save the wilderness from its saviors. Socialist arrangements have a habit of cutting both ways. When the regulatory shoe is on his own foot, however, the wildernist finds regulation to be ugly.

A rather unexpected wilderness constituency has come into being by the work of such authors as Colin Fletcher, with a large impact on the vote count in these showdowns: the armchair advocate. These people seldom or never use wilder-ness themselves, and some have never even seen an official Wilderness. They find peace of mind in the symbolism, in simply knowing that wild places are still preserved out there somewhere in the great world.

REGULATIONISM

In the early 1970s, **regulation** joined wilderness exclusions as a major threat to American industry, both on federal and private lands. Regulation, however, does not produce the instant dramatic changes that wilderness withdrawals do. It quietly makes everyday life more complicated and ultimately more expensive for the whole nation. Restrictions on harvest-ing methods such as clearcutting, on petroleum drilling methods, on the use of pesticides on farmlands, on manufac-turing industries concerning air and water pollution, occupa-tional health and safety factors, and many others add costs and lower the productivity of industrial enterprises. In rare in-stances, such as regulations that entail recovery of useful materials from wasteful processes, the benefits are real. In

14

most cases, however, the benefits are questionable: The Occupational Safety and Health Administration's vast bureaucracy has produced *no* detectable reduction in industrial injuries or accidents—it's simply costly and worthless government interference with free enterprise.

Many victims of regulation are socially invisible. For example, when the forest products industry operating on federal lands is required to use "landscape management" methods to protect "viewsheds" by shaping clearcuts as if they were natural openings in the forest, we reduce the social "cost" to hikers and tourists who want "natural"-looking scenery, but in cases where the added labor costs can be directly passed on to the consumer, we never get an accounting of the social costs to the thousands of hopeful young home-buying couples who are forced out of the market by spiraling prices.

In trying to fight the growing Regulation State, American industry was bucking a major historical trend. Since the first federal water pollution law was passed in 1948, industry has pushed for incentives rather than controls, and where controls were inevitable, for flexible rather than rigid ones. In one strangling loss after another, American industry has seen Congress give itself authority to regulate air, water, noise, health and safety, wildlife, timber harvest, and dozens of other factors. With the passage of the Resources Conservation and Recovery Act and the Toxic Substances Control Act of 1976 and the Comprehensive Environmental Response, Compensation, and Liability Act of 1980 (CERCLA, also known as the "Superfund"), all the loopholes had been closed: Congress had authority to regulate literally every substance in existence.

While wilderness issues have been fought primarily in Congress, a substantial part of the regulatory battle has been fought in the courts. In a chart prepared by the editors of *The Wilson Quarterly/summer 1977*, industry lost 7 of the 8 cases selected as landmark decisions. Observers who have noted that the industry spends a great deal of money both lobbying and litigating conclude that the dollars of industry are heavily discounted while those of the environmentalists are highly inflated when measured by the results.

Regulatory disputes have not been confined to federal

lands, either. Since the 1949 U.S. Supreme Court decision affirming the constitutionality of state regulation of private land, industrial forests have come in for a heavy burden of regulation as well.

The impact of regulation on American industry and its consuming public is enormous but poorly understood. Consider: to begin with, the metals, trees and petroleum kept from use by restrictive regulations are essential to our civilization. Not just helpful, and not just enjoyable. Essential. Of the 99 sectors listed in the *Scientific American* chart of the input-output structure of the United States economy, all 99 are affected by metals, timber and paper, and petroleum; forest products alone affect all but 11 sectors, or over 85 percent of the total economy. The infrastructure of our knowledge-based "post-industrial" society would collapse instantly without the simple commodity of paper.

Without the forest products industry, we would not miss just the obvious things we get from trees, such as lumber and plywood. Communications would have poorer eyesight and hearing without trees: photographic film and recording tape are made from wood chemicals, but who thinks of them as forest products? Fabrics (rayons and acetates), flavorings (vanillin), fragrances (the versatile turpenes enhance a huge range of products from cosmetics to cleansers), foods (maple syrup and stabilizers for mayonnaise, ice cream and other foods), and thousands of other products ranging from explosives to medicines begin in our forests.

THE CUMULATIVE IMPACT OF REGULATIONS

Forest regulations themselves take many forms: you can't harvest timber here because it's an archaeological site (two arrowheads were found nearby); there are spotted owls in this forest and you must leave a thousand acres of standing timber for each known mating pair; logging technology cannot get the timber across this creek without dropping bark fragments into the water, so it's off-limits; a local tribe uses this area as a Native American Religious Site, so you can't cut trees nearby; the costs of roadbuilding into this timber are too high to justify logging; the elevation and soil type here are too risky for growing new trees so you can't cut the ones that are here now; there has to be a buffer strip of standing trees left for 200 feet on each side of this stream to keep the solar loading factor from raising water temperatures too high for salmon spawning.

More than 200 *different kinds* of restrictions add up to thousands of specific restrictions on U.S. Forest Service lands alone. Each of these non-timber designations no doubt has a legitimate purpose. But an endless stream of annually increasing restrictive regulations means less land in commercial forests means declining supply in the face of steady (or rising) demand means sharply higher prices (ask any home buyer or office supply house), for Congress has not yet managed to repeal the law of supply and demand.

The impact of regulation can be pernicious. Increased red tape, labor, and time always mean higher internal costs, but these costs cannot always be passed on to the consumer. Wood, remember, is a commodity. One piece (of a given grade) is much like any other piece, and pure price competition prevails over any other market force. Commodity brokers will simply go to the best price available. Timber firms must therefore frequently eat the higher costs from regulation, which means lower profits and dividends, less ability to form capital and expand, or in some cases even to remain competitive.

Worst of all, once Congress has given some bureaucracy a regulatory mandate in law, environmental lobbyists find it easy to pressure the buraucracy to constantly tighten regulatory provisions and to increase the gross number of regulations—and Congress, having gone on to other matters, isn't even aware of the devastation those cancerous regulations wreak in logging communities as one mill after another goes under. Then, adding insult to injury, environmental lobbyists say, "Oh, it wasn't environmental regulations that did it, it was just a shift in the economy caused by other forces." When you're squeezed out of the market because of regulatory restrictions on supply, your business is not only more sensitive to such economic forces as interest rates, but your economic sector also *influences* them by pushing investment capital into other sectors and other countries.

Perhaps most ominous for the future of regulation, certain

17

leaders of the environmental movement see themselves as the vanguard for a "new society." The new environmental society is not clearly defined in environmentalists' own minds, but it obviously cannot be realized in a world structured by industrial capitalism, as pointed out by historian Donald Worster in *The Journal of Forest History* of January, 1986. For many years I have made this assertion, which has been greeted by derisive denials of environmentalists. However, in 1984, Prof. Lester W. Milbrath of the State University of New York published *Environmentalists: Vanguard for a New Society*. His book and his passionate advocacy for an environmental revolution to change the basic economic structure of America is the best clincher my argument could have.

It is time to look under the hood of this engine of social destruction, for that is exactly what the environmental movement has become, before it does irreparable harm to *both* our economic system *and* our natural heritage. The stakes are high. The outcome will affect the work, the purchasing power, the play, and the general peace of mind of every living American, as we shall see.

WHY ECOLOGY WARS?

Conflict over the environment is nothing new. As a little reading in American history will reveal, the fundamental disagreements between environmentalists and industry have been smouldering for well over a hundred years. They flared up during Teddy Roosevelt's era as The Conservation Movement, a landmark coming-to-grips with the new reality that the American frontier had closed, that the nation faced an age of increasing land scarcity, and therefore, increasing resource scarcity. The Rough Rider himself declared in 1905 that the object of land policy was "to consider how best to combine use with preservation." But the environmental movement of today, with its new twists of ecological awareness, its land ethic, and its reliance on big government to enact and enforce its programs, has blossomed only in the last two decades. In the span of a few short years during the mid-1960s and early '70s, a sizeable portion of the public's whim of steel swung away from its traditional focus on material well-being and toward the blue-jeaned counterculture's urge to throw out the industrial baby with the polluted bathwater and go back to Nature, whatever that was conceived to be.

These great shifts in our society have for the most part baffled the leaders of American industry. I've heard many an angry manager ask, "Why do these environmentalists want to destroy our free enterprise system?" And it's not just the leaders who are worried. I've heard down-home miners and loggers ask, "Are they communists?" I've heard ranchers and farmers ask, "What do they want?" But even industry's best experts can't seem to get a grip on environmentalism. While industrial leaders have developed excellent material technology, they have suffered absolutely terrible public relations, and in the political arena were push comes to shove, they've been losing hands down.

Part of the problem is that American industry is just too busy minding the store to keep up on fast-changing issues. Associations such as the National Association of Manufacturers, the American Petroleum Refiners Association, the National Cattlemens Association, the National Forest Products Association, the National Mining Congress, the National Agricultural Chemical Association, the National Ocean Industries Association, and enough others to fill a dozen pages in the District of Columbia telephone directory, all struggle valiantly to put out one environmental brush fire after another. In the pandemonium, they often chart unworkable strategy, as measured in the vote count in Congress and decisions from U.S. Circuit Courts of Appeal and the U.S. Supreme Court. A significant part of this failure, in my assessment, lies in the simple fact that American industry does not understand environmentalism, and therefore does not grasp the true nature of the conflict.

There are, of course, other powerful reasons why American industry has few friends in the general public and is losing many of its political battles. For one thing, industry has few friends in the media. As Mobil Oil vice president for public affairs Herbert Schmertz wrote in a 1984 issue of *Washington Journalism Review*, many "newspersons see crime in the suites. This is not hyperbole. Morley Safer of '60 Minutes' said that, 'No businessman who has made a success for himself is entirely clean, probably.' And, according to the non-profit Media Institute, 'almost half of all work activities performed by businessmen' on prime-time TV series 'involve illegal acts.' The report added that 'Television almost never portrays business as a socially useful or economically productive activity.'" In fact, the media, both electronic and print, display sufficiently regular hostility to free enterprise that citizen watchdog organizations have arisen to combat the problem, such as Reed Irvine's Accuracy In Media (AIM). Media power is a substantial factor in industry public relations.

For another thing, as Daniel Bell pointed out in his *The Coming of Post-Industrial Society*, the largest part of our employment since 1955 has been increasingly in the service sector. Why is this important to ecology wars? Because it means that most Americans are thereby comfortably buffered from and unaware of the hurly-burly realities of basic resource extraction and conversion in our mines, forests, rangelands, oceans, farms and factories. The total goods sector tends to appear only as a source of pollution to those in the service sector when it is visible at all. This out-of-sight, out-of-mind factor combines with other forces.

THE POST-MATERIALIST BLUES

The new generation of affluent young citizens now taking over the economy's reins, as noted by Ronald Inglehart in his massive study *The Silent Revolution*, was unscarred by The Great Depression and World War II. They have never known dire want or monstrous physical insecurity. These "post-materialists," as Inglehart dubs them, are less obsessed with success and security than their parents were. They are more oriented toward personal autonomy, needs for love and a sense of belonging, and intellectual and aesthetic pursuits. They do not fully realize how crucial the well-being of industry is to their own well-being, and haven't the faintest notion of industry's needs. They tend to despise crass commercialism, and are politically very liberal and very active. This New American Society, a large minority consisting of perhaps 25 percent of the total population of this country, forms a ready-made base of support for environmentalist causes, and a potent one. We will have more to say about "post-materialists" later.

20

The two major philosophical views about land use planning are *the free enterprise approach* that relies on markets and enlightened self-interest to allocate scarce resources and *the statist approach* that relies on government control in one form or another to determine land uses.

The major disputes in the forest environmental battle are those that pit private commercial forest firms using market-oriented management practices against governmental restrictions on commercial use. Forest conflict has centered around the two major fronts of our ecology wars: wilderness preservation (removal of federal forests from commodity use), and regulation of commercial forests by legislative and bureaucratic fiat. These two issues of wilderness and regulation are so complex that they can be understood only when viewed in their historical contexts.

ROOTS OF CONFLICT

At the dawn of the new environmental era, Congress, in The Multiple Use-Sustained Yield Act of 1960, directed the Forest Service to give equal consideration to outdoor recreation, range, timber, watershed, wildlife and fisheries resources. The idea was to provide something for everyone. By and large, foresters throughout the industry approved of the new policy with such responses as "We've been practicing multiple use all along."

Multiple use is basically a simple idea, to give the greatest good to the greatest number, but it is exceedingly complicated to put into practice. And it is most definitely and emphatically a statist approach: Multiple use applies only to government forests, and the mere existence of government forests is itself a completely socialist arrangement, although we seldom think of it as such. Sound far out to call American national forests *socialist?* Think it through. The essence of socialism is government ownership of the means of production—virtually every thinking person will grant that basic definition. Land is the basic means of *every kind of production* and land is obviously the basic means of production of forests, as we noted above. The U.S. government owns one-third of the total land area of America. It's true. A third of our nation, including a substantial fraction of our forests, is held

21

in a socialist arrangement. We seem content to let things be that way, but the fact should be kept in mind every time we think of environmental conflict.

Similarly, sustained yield, the concept of cutting no more timber in a given year than grows back in the same time so that you never run out of trees, is easy to say but complex to administer in the forest. Sustained yield is less an inherently statist approach to forest management than multiple use, because private firms may wish to manage their own fee-title timberlands under sustained yield principles. But within the firmly rooted socialist framework of our national forests, back in 1960 sustained yield as a government policy made good sense to the industry and became its major rallying cry in the ecology wars that were to follow.

Wilderness, at first blush, has a wholesome ring in modern America, and it is one of history's perversities that such horrendous controversy has arisen over it. A good part of the reason for such controversy is that Wilderness (capital W) is not wilderness (small w). A 1978 survey by Opinion Research, Inc., found that more than 75 percent of all Americans still didn't realize the difference between just any woodsy recreation spot and officially designated wilderness. The difference is monumental. The Wilderness Act of 1964, which created the National Wilderness Preservation System, mandated that Wilderness is an area of at least 5,000 federally-owned acres and defined it thus:

"A Wilderness, in contrast to those areas where man and his works dominate the landscape, is hereby recognized as an area where the earth and its community of life are untrammeled by man, where man himself is a visitor who does not remain."

In practical terms, that means no roads, no buildings, no motorized vehicles, no timber harvest, no watershed management, severely restricted fire, insect, disease, and wildlife management, and in most places not even toilets. It's the law. But the average American never even heard of it.

To many industries this new law amounted to institutionalizing a confiscatory single use on federal lands in violation of the intent of the earlier Multiple Use - Sustained Yield Act. However, when institutions are built around socialist

22

arrangements such as federal lands, this type of rude political shock is to be expected. To the successful conservation movement, led by the Wilderness Society and Sierra Club, it was a triumph of ecological sensitivity over "Multiple Abuse," as they had come to call timber harvest and other extractive uses of forest land such as mining and petroleum production.

The passage of the Wilderness Act of 1964 marked the watershed in thinking about the American forest, and resource industries have never really been able to cope with it. In one Wilderness showdown after another—the Alpine Lakes Wilderness Area, the Mineral King Recreation Development, RARE I (Roadless Area Review and Evaluation process involving 56 million acres of inventory and 12 million acres of study areas), the Endangered Wilderness Act of 1977, and various Wilderness packages of the 1980s—American industry has gone down to defeat in Congress and the courts. Industry has lost three out of four of its congressional lobbying campaigns since 1960. The great lobbying power of the "Timber Barons" and the "Oil Monopoly" and the "Mining Kings" is one of the more ironic of our quaint public myths today. Businesses, because they exist for profit, are immediately suspect in the lobbies of Congress, regardless of the merit of their argument. Environmental lobbyists are seen by legislators and the bureaucracy as proponents of the common good regardless of the absurdity of their argument.

This pattern is merely the new form of an old struggle, for since John Muir fought for the establishment of Yosemite National Park in the 1880s and the Forest Service in 1924 set aside the world's first official Wilderness Area in New Mexico's Gila National Forest, environmentalists have successfully lobbied more than 77 million acres into the National Park Service's preservation programs and more than 80 million acres into Wilderness (or similar "Primitive Area" designations). By 1985, more than 5 percent of America's total land area (nearly twice as much as is owned by the forest industry) had been set aside in some kind of federally-controlled noncommercial status. The figure is climbing every year as Congress steadily eats away at our basic resource areas. If every study and proposal now on the books were to be taken into account, just to give you an idea of the grandeur of the wilderness mystique, restricted use of some kind is envisioned for *all* federal areas, which, remember, occupy *one-third of America's land.*

All these factors affect employment levels, and we should not forget that the combined Lumber & Plywood and Pulp & Paper sectors feed more than $65 billion annually into our economy and their 1.2 million workers represent 6.6 percent of all U.S. manufacturing employment. However, contrast this with the 4.5 million members of the National Wildlife Federation alone out of the top 10 environmental groups, and you will see how overwhelmed a single industry in modern America can get in the political arena. It is obvious that American industry needs a grass-roots citizen support movement like the environmental movement if our economy is to regain its full vigor and grow at proper rates again.

There's no denying that a new member has come to America's power elite: The environmental establishment. Multi-million-dollar non-profit environmental organizations are quite a force unto themselves, but the federal environmental bureaucracy entrenched in such agencies as the Environmental Protection Agency, the President's Council on Environmental Quality, the Department of the Interior, and Department of Agriculture, among many others, commands power that is potentially crushing.

GOVERNMENT'S BATTLE FRONT

Government today forms a very significant third corner complicating the industry/environmentalist contest. Not only does government make the basic rules of the natural resource game, it also owns a sizeable chunk of the playing field. The federal government alone owns 762 million of the 2.27 billion acre total land area of the United States—and that's not counting state, county and municipal land. As I've repeatedly noted, the feds own one-third of the entire nation, or an area equal to all America's forested land. Government is supposed to act as a mediator on behalf of all its citizens, but too frequently slips into an adversary third-party role with priorities of its own. Then the age-old question arises, "Who will guard the guards?"

The late master strategist and futurologist Herman Kahn of

the prestigious Hudson Institute put his finger on this problem when he warned of the "Health and Safety Fascists." This element, Kahn wrote, "has been singularly unreflective about its advocacy of strict, no-compromise controls on business and the environment regarding health, safety, and ecological matters. It has been unaware of its irrational prejudice against both business and the middle class, and consequently of the extent to which its regulatory zeal, from the viewpoint of both those it wants to regulate and of the general public, smacks of the dictatorial."

ECOLOGY WARS ARE HERE TO STAY

Some analysts feel that we are witnessing in the 1980s the maturity of the environmental movement, but I don't think so. I believe that what we have seen is only the first phase, the setting in place of all the laws and court decisions. In the next decade or two we are likely to see several things, first, the screws of wilderness withdrawal and regulation tightening a little here, a little there, despite the legacy of the Reagan administration. Then the revolutionary efforts of the "Vanguard for a New Society" in the form of eco-terrorism will become more widespread. A political agenda will appear that tries to strip our federal lands of *all* extractive resource use, ranging from livestock grazing to petroleum drilling. Then the "Vanguard" will attempt a genuine political takeover, through the ballot box and Political Action Committees if possible and by direct action, to use Professor Milbrath's words, if the democratic process fails.

The day will come when the flexibility of our market economy can no longer bear the strain, and our open society no longer survive in its present form. We may well see the day when regulation grows so vast that totalitarian measures seem both necessary and acceptable to a large public. If this seems too melodramatic, read ecologist William Ophuls' *Ecology and the Politics of Scarcity*, which maintains that modern civilization has outlived its usefulness and must be governed by "implacable ecological imperatives."

American industry has a moral obligation to protect itself from environmentalist attacks. All industry must come to grips with environmentalism, learn what it is, what motivates

25

its actions, what the shape of its history is, how its propaganda works, how to combat it to the fullest extent possible. We have only touched lightly on these items in this first chapter, but each point will subsequently be clarified in detail.

This is the message America must get: As crass as business tends to be, it is still an essential part of the whole, a vital part of the human ecosystem. And, as sensitive as the environmental movement tends to be, it is now in a position to wield colossally blundering economic power. It was Oscar Wilde who defined a cynic as one who knows the price of everything but the value of nothing. We must be warned that the environmentalist is one who knows the value of everything, but the price of nothing.

2

Assessing Environmentalism

HOW DO YOU MAKE SENSE of environmentalism? The unhappy fact is that many in American industry don't, and the political costs are unbearable.

Yet environmentalism does have a rationale, and it can be understood, and I don't mean in the bandwagon way that the public has embraced it; I mean in the strategic sense.

The converging social and political pressures of environmentalism make the central question before American industry, "Will we be allowed to do business within a decade or two?" Therefore, understanding environmentalism is crucial to our whole economy, our whole nation; doing something effective about it is essential.

If American industry is ever going to form a more effective check and balance against its opposition (recall that it's now losing three out of four legislative battles to environmentalists), it must first perform a methodical analysis and assessment of environmentalism, preferably from an insider's viewpoint. Having lived and breathed environmentalism for a decade before the subject became fashionable, I'm prepared to offer that insider's view for the benefit of vital industries that I have come to know and respect.

INDUSTRY'S IDEOLOGY-PHOBIA

There is a built-in difficulty in assessing environmentalism: industry leaders don't mind examining the specific issues of one controversy after another, but they shy away from analyzing the underlying ideology. This reluctance to skate on the thin ice of ideology is understandable, but it puts American

27

industry in the same predicament as a doctor who carefully studies the obvious symptoms of a disease without ever looking into the obscure causes.

Any understanding of environmentalism worth having requires an intellectual effort far more extensive and subtle than a reflex action against the threat to our earnings. I know that this is likely to make the bottom line-oriented manager impatient, and irritate the down-to-earth logger, miner, or steelworker. But if a successful opponent is neither down to earth nor bottom line-oriented, we'd better find out why not, and quick.

I've used the best available scientific studies of ideology and social conflict so the assessment will be as objective as possible.

Before we can dig into the ideology of environmentalism, we have to clear up a point of business ideology. I'm assuming that there's more to running a one-employee small business or a megabuck corporate giant than the economic purists like Milton Friedman would have it, that the only legitimate purpose of a corporation doing business is to meet a demand and make a profit. No, it seems to me that economist Neil H. Jacoby's model is more to the point, that corporations must respond to both market forces and non-market forces like environmentalism, because both affect your firm's costs, revenues, and profits. It may be heresy, but brother, it's real.

The usual way to begin an assessment is to define the problem, and that requires us to define environmentalism. What is environmentalism? Hundreds of workers and managers have asked me that question. Everyone seems to have an answer, as H. L. Mencken quipped, easy, simple—and wrong! I can assure you that there is no simple answer.

THE ANATOMY OF ENVIRONMENTALISM

But there is a correct answer. However, defining environmentalism is the wrong approach. Just as a physician does not "define" the human body with mere word play, but rather examines its anatomy for better structural and functional understanding, so our best entry point into environmentalism is to take it apart piece by piece to see what it's made of and how it works.

The environmental movement of the last two decades is an amalgam, a crazy quilt of fifteen or more separate historical trends that got welded together during the 1960s, some as much by accident as by intention. These trends are:

1. The older conservation movement from Teddy Roosevelt's era, more militant in tone, perhaps, but still centered on wilderness preservation and wildlife protection.

2. The anti-pollution movement that grew from public outcry against the "killer smogs" in Donora, Pennsylvania in 1949; the nuclear fallout scare of the mid-50s with its screaming headlines "Strontium-90 in Babies' Milk!"; the pesticide bugaboo dramatized in 1962 by Rachel Carson's *Silent Spring*; the plight of our garbage-laden oceans as denounced in a 1966 National Geographic television special, *World of Jacques Cousteau*; oil-soaked birds from a seafloor wellhead blowout in the Santa Barbara Channel; and Barry Commoner's punching of the phrase "environmental crisis" in his 1971 *The Closing Circle*. This movement brought a cloud of alarmism over America which terrified people with claims that all life on earth was in imminent danger.

3. The nature appreciation trend that is taken for granted today, but which grew into a major cultural feature of western society only after a centuries-long struggle with a religious bias against wilderness, and a huge accumulation of pro-nature literature. Today the nature appreciation trend is reflected on every book publisher's list, in magazines such as *National Geographic* and *International Wildlife*, in the most powerful newspapers such as the *Washington Post*, in motion pictures and on television.

4. The outdoor recreation boom resulting from growing affluence and mobility in the general population, and reinforced by the "cult of the simple rustic life" that pervades our advertising (Marlboro and Salem cigarette ads in particular), that supports hundreds of national magazines on camping, boating, hunting and fishing, climbing and hiking, and that glorifies life in the suburbs and rural areas. One of the biggest tactical errors of American industry has been to confuse the recreation boom with the nature appreciation trend, a problem I will detail later.

5. New life attitudes and values based on the science of ecology with its findings that all life on earth is interrelated, and the "ecological conscience" that holds the environment to be a community to which humans belong, not a commodity that we possess. If the environment, particularly land, were to be legally removed from its status as "a commodity that we possess," there would be severe but unmentioned consequences upon the American way of life. The impact of the "ecological conscience" if politically realized would be devastating to private property rights and capitalism. The "ecological conscience" and free enterprise cannot co-exist. However, this inherent, unavoidable anti-capitalist strain in environmentalism is seldom discussed openly by environmentalists.

6. An anti-hunting, anti-gun movement advocating "animal rights." The anti-gun movement is described in detail by Alan M. Gottlieb in *The Gun Grabbers* published by Merril Press. The anti-hunting crowd has no regard for Second Amendment rights and wants to see all guns banned, both handguns and long guns.

7. A counterculture protest against established American institutions and values, and for "personal freedom," long hair and beards, folk music (including John Denver and Gordon Lightfoot with their "ecology songs"), drugs, sexual liberation, and mystical religions, particularly the nature religions of the Orient (Jainism and Taoism), and of the American Indian (popularized by Carlos Castaneda's *Don Juan* series, *Black Elk Speaks*, and other books). As Daniel Bell remarked in *The Cultural Contradictions of Capitalism*, "The cultural impulses of the 1960s, like the political radicalism which paralleled it, are, for the while, largely spent. The counterculture proved to be a conceit. It was an effort, largely a product of the youth movement, to transform a liberal life-style into a world of immediate gratification and exhibitionistic display. In the end, it produced little culture and countered nothing."

DETECTING SPIRITUAL CRAP

8. Perhaps more lasting than the counterculture itself, which is now remembered mainly by a few mildly interesting psychedelic posters and some very nice music by the Beatles, was the broader "New Age" life philosophy. New Age figures such as Werner Erhardt, Ram Dass, Uri Geller, and Arthur (*Primal Scream*) Janov got people interested in "human potential," the Tao of this and that, new physics, consciousness, philosophy, *Zen and the Art of Motorcycle Maintenance*, and other esoterica. Mark Satin's *New Age Politics: Healing Self and Society*, a book symptomatic of the movement, envisioned the New Age as neither right nor left, but instead interested in finding appropriate solutions to problems—not simply alternative methods; able to reconcile people to each other's needs; concerned with the specific ethics and political values that will permit everyone to survive, grow and flourish; and equally concerned with the personal and the planetary.

Michael Rossman, however, tuned his "spiritual crap detector" to these lofty New Age ideas and commented wryly in *New Age Blues* on the fact that people re-create authoritarian social systems when attempting deep social change. Rossman also looked at the negative side of positive thinking (if you think problems are really just opportunities, somehow your opportunities never get solved).

Perhaps the most successful New Age book was *The Aquarian Conspiracy: Personal and Social Transformation in the 1980s* by Marilyn Ferguson. The dust jacket blurb read, "A leaderless but powerful network is working to bring about radical change in the United States. Its members have broken with certain key elements of Western thought, and they may even have broken continuity with history. . ." Ferguson was not talking about Congress, she was talking about New Agers plying the paranormal hinges of our society trying to make its doors open in new directions. The book is an immensely entertaining compendium of New Age enthusiasm for the left-brain/right-brain theory, networking, all about paradigms and paradigm shifts, and other ineluctable goodies. People take this sort of thing seriously and bring these mindstyles to environmental conflicts, which means it pays to be forewarned and fore-informed about them.

NOT GROUCHO

9. A popularization of Marxism and revolutionary chic that went along with the counterculture and the New Age movement. There is a great deal more to Marxism than the average American knows. Most Americans have not read a single one of Karl Marx's dozens of books or hundreds of articles and letters—not surprising since he is such a crashing bore—which leaves us shamefully ignorant and vulnerable as a culture. During the late 1960s, a distinctively Marxist-Leninist strain of anti-capitalism crept into the environmental debate, particularly in the media. It was reflected in such comments as "in order to protect the environment we must eliminate private property," and "the word 'ecology' implies the indivisibility of total systems, and therefore environmental problems are not susceptible of separate solution or even reform but imply structural attack on the political and economic system itself"—which was the message of Alexander Cockburn and James Ridgeway's 1979 collection of essays entitled *Political Ecology.*

Perhaps most telling, Soviet publications during the 1970s glorified certain American environmentalists. For example, Grigori Khozin in *The Biosphere and Politics,* heaped praise upon Professor Lynton K. Caldwell, one of the two chief architects of our National Environmental Policy Act of 1969 (the other being the late Senator Henry M. Jackson). Khozin was ecstatic about Caldwell because the professor, while writing a strictly analytical passage in his 1972 book *In Defense of Earth,* had likened the environmental movement to Marxism. Khozin gushed: "Lynton Caldwell, a prominent American student of international cooperation in environmental protection, has likened this major movement of our times—whose purpose is to find rational forms of controlling the interaction of human society and nature—to Marxism, whose authority in the world is growing. In Caldwell's opinion, the environmental movement, just as Marxism, represents an 'action-oriented philosophy; it claims . . . a base in science; it has developed a set of propositions regarding man's relationship to history and the existential world and it leads to a specific course of action.' "

In fact, our American colleague was not endorsing Marxism at all, although Comrade Khozin would undoubtedly like us to think he was. Environmentalists are deeply suspicious of Marxism because Marx envisioned communism as the ultimate conquest of nature by man's technology (in this regard Marx is closer to Western capitalism than environmentalists are). Fortunately, Khozin's attempt to co-opt the American environmental movement is so transparent as to be laughable. No doubt Professor Caldwell is embarrassed at this shameless propaganda use made by the Soviets of his academic analysis of environmental activism.

This sort of Soviet intellectual thuggery can lead to real problems when unwitting American journalists swallow the ideas whole and begin to disinform the rest of us with "factoids"—facts that seem to be factual but are in fact not factual—in seemingly sound journalism. Many "America-bashing" articles in our own nation's news clippings are centered around the theme of capitalist evils destroying the environment, which is the main theme of a whole series of Soviet publications (see the bibliography entries under Budyko, 1980; Dioumoulen, et al., 1983; Pavlenko, 1983; Kortunov, 1979; and Novik, 1981).

It is obvious that industrialization has no special dispensation to be clean or dirty just because it is communist or capitalist. Americans would do well to familiarize themselves with Karl Marx's more accessible works such as *The Poverty of Philosophy*, *The Economic and Philosophic Manuscripts of 1844*, *Wage Labor and Capital* and *The Communist Manifesto*. It would not hurt us to also become familiar with some of Vladimir I. Lenin's works such as *What Is To Be Done?*, *Imperialism, the Highest Stage of Capitalism* and *Marxism on the State* in order to recognize Marxist-Leninist ideas in the environmental debate when we see them and to avoid being blind-sided by Communist efforts to co-opt the environmental movement.

Familiarizing ourselves with Marxism-Leninism is particularly important at this juncture because Marxists are openly appealing for environmentalists to "join the ranks." David Pepper, a self-described Marxist, invites environmentalists to adopt Marxism as their guiding philosophy in his 1984 book *Roots of Modern Environmentalism*. According to Pepper, the only realistic, effective strategy is for environmentalists to join the socialists' campaign for revolution and world justice. In a Marxist society, Pepper assures us, people would be less materialistic, less emulative, and less alienated from nature. Pepper's arguments are considerably weakened by a reading of Marx however, since the patron saint of Communism constantly advocated the total conquest of nature through technology and human reason. Modern Marxists such as the Neo-Hegelians have pounced upon one small reference in Marx's *Economic and Philosophic Manuscripts of 1844* to "respect for nature" as a basis for the construction of an "eco-socialism" that might appeal to modern environmentalists. Lester Milbrath even suggested in his book *Environmentalists: Vanguard for a New Society* that "Some day we may see a merger of the New Environmental Paradigm [Milbrath's name for the lifestyle/mindstyle of environmental reformers] with a new 'Ecological Marxism.'"

THE DARK SIDE OF EVERYTHING

10. An awakening to the finite resources and limited population carrying capacity of the earth as trumpeted by Donella Meadows, et al., in their pessimistic *The Limits to Growth*, by Paul Ehrlich's *The Population Bomb*, and ironically reinforced by beautiful NASA photographs of earth from space showing it to be small, alone and precious—"Spaceship Earth."

11. The consumerism movement as founded in 1936 by Consumers Union and capitalized upon in the '60s by former employee Ralph Nader, especially as it relates to the energy crisis and to "health foods" without chemical additives or pesticide residues, and reflecting awareness of man's position on top of the food chain.

12. An intellectual anti-technology cult that asserts that our machines have become our masters, forcing us to do tedious and degrading work, to consume things that we do not really want, and that a technocratic elite is taking over control of society, as preached by Siegfried Giedion's *Mechanization Takes Command*, Jacques Ellul's *The Technological Society*, Lewis Mumford's *The Myth of the Machine*, and Herbert Marcuse's Marxist analysis, *One-Dimensional Man*.

13. An anti-civilization trend insisting that man's basic nature is thwarted by the constraints of civilized living. Sigmund Freud examined this idea in his 1930 *Civilization and its*

Discontents, but it was brought to the fore in the '60s and '70s with suggested remedies like simply leaving civilization to live off the land (*The Survivalists* by Patrick Rivers) and such bizarrely grandiose projects as totally dismantling industrial civilization (*Person/Planet, the Creative Disintegration of Industrial Society* by Theodore Roszak).

14. A dark strain of anti-humanity that despises everything human, wide-spread in environmentalist writing, but particularly influential in the poetry of Robinson Jeffers, whose poem "The Answer" (Random House, 1926) gave the militant Friends of the Earth the name for their news bulletin *Not Man Apart* (". . . the greatest beauty is / Organic wholeness, the wholeness of life and things, the / Divine beauty of the universe. Love that, not man / Apart from that . . .") but whose underlying disgust with people was revealed in these lines from his "Original Sin" (Vintage, 1963): "As for me, I would rather/ Be a worm in a wild apple than a son of man / But we are what we are, and we might remember / Not to hate any person, for all are vicious / And not be astonished at any evil, all are deserved; / And not fear death; it is the only way to be cleansed." These poetic sentiments of revulsion at humanity have had a profound impact on idealistic young students, and have particularly baffled workers during confrontations, because they could not identify, much less believe, such irrational feelings.

SELF-APPOINTED SAVIORS

15. To these major trends I would add the tendency of some analysts to put all of these trends together and to see environmentalism as a soteriology, that is, a universal doctrine of salvation based upon revelation. The revelation, of course, is not divine, but rather naturalistic: the findings of the science of ecology. The salvation, of course, is not of man's immortal soul, but of the earth as an ecosystem, and among the more generous environmentalists, of humanity as a whole (no nasty warty individuals allowed).

We saw the first evidence of this tendency in Aldo Leopold's "land ethic" from his lyrical *A Sand County Almanac* (Oxford University Press, 1949); "It is inconceivable to me that an ethical relation to land can exist without love, respect, and admiration for land, and a high regard for its value. By value, I of course mean something far broader than mere economic value; I mean value in the philosophic sense."

Leopold evidently never realized that his definition would be just as true if his word "land" were to be replaced by the phrase "American industry". Read it yourself that way and see. Ethics is a broad avenue not reserved for environmentalists only. Our society could do well to give some respect and admiration, if not love, to its benefactors in American industry, and to recognize that industry's only product is not pollution, that it embodies "value in the philosophic sense" of innovation, self-sufficiency, service, resourcefulness, stewardship, strength of character, and perseverance in the face of overwhelming obstacles.

This "land ethic" of Leopold's inspired a large following with significant effect on congressional wilderness campaigns, particularly the passage of the Wilderness Act of 1964.

16. In 1984, Lester Milbrath, a political scientist and director of the Environmental Studies Center at the State University of New York-Buffalo, came up with one of the first outright analyses of environmentalism as a revolutionary means by which to overthrow existing political arrangements in America. His book *Environmentalists: Vanguard for a New Society* passionately identifies environmentalists as the "vanguard" of a "revolutionary but peaceful social change" leading to a society which seeks "a better way of life in a long-run sustainable relationship with nature" by promoting renewable energy, avoidance of physical risks in the production of wealth, and opportunities for citizens to have a say in public policies, among other things. Environmentalists in America, England and West Germany, according to mail surveys carried out by Milbrath between 1980 and 1982, share a world view which he calls the New Environmental Paradigm (NEP). This NEP includes, as Milbrath puts it in a chart:

I. A high valuation of nature: A. Nature for its own sake (worshipful love of nature); B. Humans harmonious with nature; C. Environmental protection over economic growth.

II. Generalized compassion toward: A. Other species; B. Other peoples; C. Other generations

III. Careful planning and acting to avoid risk: A. Science and technology not always good; B. No further development of nuclear power; C. Development and use of soft technology; D. Regulation to protect nature and humans—government responsibility.

IV. Limits to growth: A. Resource shortages; B. Population explosion—limits needed; C. Conservation.

V. Completely new society (new paradigm): A. Humans seriously damaging nature and themselves; B. Openness and participation; C. Emphasis on public goods; D. Cooperation; E. Post-Materialism; F. Simple lifestyles; G. Emphasis on worker satisfaction in jobs.

VI. New politics: A. Consultative and participatory; B. Partisan dispute over human relationship to nature; C. Willingness to use direct action; D. Emphasis on foresight and planning.

Although this NEP (New Environmental Paradigm) is somewhat vague about how the new society will actually feed, clothe and shelter itself, evidently a substantial number of Americans believe in it. Milbrath claims that a rearguard opposing the NEP clings to a Dominant Social Paradigm (DSP), which is the flip side of the NEP and which dominated western societies until the 1970s.

Milbrath's NEP is tiresomely reminiscent of all the Utopias in history and unconsciously or otherwise employs Lenin's notion of a "vanguard" leading the masses. Instead of Lenin's Communist Party as the vanguard of the proletariat, Milbrath gives us the Environmental Movement as the vanguard of the service sector. Milbrath's concept that the NEP can be placed at the vanguard of a social spectrum with the DSP at the rearguard is the typical we-versus-them, two-dimensional thinking that environmentalists so often fall into. Everyone in society, Milbrath claims, can be arrayed at some point on the straight line between these two extremes. It would never occur to Milbrath that both his Dominant Social Paradigm and his New Environmental Paradigm are out of date and are being superceded in mid-1980s America, Britain and West Germany by a Compatible Disequilibrium Paradigm which assumes and asserts that humans in all their variations and nature in all its variations form a dynamic system that represents a new level of evolution, and that people cannot be accurately arrayed on any single axis, but must be evaluated along many dimensions and from many angles. We will discuss this new paradigm further in Chapter Seven, **Defeating Environmentalism.** Here, then, is the gross anatomy of environmentalism. There may be more to it, but there is certainly nothing less.

WHY DID ENVIRONMENTALISM HAPPEN?

These varied forces, all of them important, came together in the years surrounding the media blitz of Earth Day, April 22, 1970, the peak of the environmental movement's popularity in America. Obviously, environmentalism is no simple movement of folks out to cop some big acreage for Sunday walks in the woods.

There is a quarrel among scholars over whether ideas or economic forces caused the growth of environmentalism. Marx felt that the "economic base"—how we make a living, whether in primitive communist tribal systems, slave-based societies such as Rome, feudalism, capitalism, socialism or "full communism"—determines the "social superstructure" of ideas and institutions. Thus, a Marxist would look to the "economic base" to explain how it gave rise to a "social superstructure" institution such as environmentalism. Soviet ecologists such as Mikhail Budyko in *Global Ecology* say that "environmental degradation is an inevitable consequence of capitalist societies"—as if communist factories did not pollute—and the "ecological crisis of capitalism" gave rise to environmentalism, a modern echo of Marx's antique concept that boom-bust crises would eventually bring capitalism to a natural self-destruction. On the other hand, liberal historians such as Roderick Nash in *Wilderness and the American Mind* feel that attitudes and ideas, not economic formations, shaped today's environmentalist trend.

I say it's both. From my own theoretical suppositions, there is no reason to assume the economic-base / social superstructure dichotomy is realistic. But even granting it for the sake of argument, the economic base and the social superstructure are certainly interactive and influence each other; neither completely determines the other, but both influence the other. Environmentalism may be a social ideology, but it did not grow in an economic vacuum. Dynamic socioeconomic forces made it possible for the ideology to arise and flourish. But what were those forces?

SORTING OUT SOCIAL FORCES

Most important was the coming of "post-industrial" society with its labor-force majority in service occupations, its emphasis on science and information, and its better educated population, all of which fertilized the soil in which environmentalism could bloom.

Changing values and increasing political skills were other important forces, as laid out by Ronald Inglehart in his magisterial *The Silent Revolution: Changing Values and Political Styles Among Western Publics*. Inglehart's tremendously important work established that "The values of Western publics have been shifting from an overwhelming emphasis on material well-being and physical security toward greater emphasis on the quality of life." A significant shift is also taking place, says Inglehart, in the distribution of political skills.

Four important "system-level" changes are taking place in our society, according to Inglehart: 1. Economic and technological development is bringing satisfaction of basic sustenance needs to an increasingly large proportion of the population. 2. Distinctive cohort experiences are giving the younger generation a new outlook on life shaped primarily by the absence of "total" war (despite the trauma of the Korean and Vietnam conflicts). 3. Rising levels of education, especially a higher proportion of the population obtaining college educations are changing society's values. 4. Expansion of mass communications and increasing geographic mobility of our society are also changing values.

These four "system-level" changes create individual-level changes. Economic development and the affluence that came with the absence of "total" war freed our minds from basic needs and led to an increasing emphasis on our individual needs for a sense of belonging, self-esteem and self-realization.

Higher education and mass communication gave the younger generation increasing skills to cope with politics on a national scale. They also changed personal values. Inglehart discovered that college life makes students more liberal, more tolerant and more likely to challenge authority. Mass communication such as television introduces dissonant signals into our homes and shows alternate lifestyles. Both higher education and mass communication make it more difficult for parents to transmit their personal values to their children in unaltered form.

In their own turn, these individual-level changes brought about further system-level changes. Political issues changed. "Lifestyle" issues became increasingly salient. The social bases of political conflict changed: elite-directed political mobilization and class conflict gave way to elite-challenging issue-oriented special interest conflicts.

The Role of Individual Needs

The historical sequence went something like this: Even though many "pre-affluent" communities of the poor, of blacks and other minorities still struggled with civil rights and basic economic issues, the security and affluence of the post-World War II era left a sizeable segment of the population—mostly in the service sector—free to climb psychologist Abraham Maslow's "needs hierarchy." They were able to satisfy their basic needs on the hierarchy's bottom rungs of food, clothing and shelter. Once people satisfy this level of needs, Maslow discovered, the basic needs no longer motivate action as strongly and a new set of needs arises.

These new higher level needs are non-material and arise in a more or less regular order. They progress upward through the needs for love, a sense of belonging, self-esteem and the need for self-realization. Some of the population was able to gratify all these needs. When these higher-level personal needs are gratified, they too no longer motivate action as strongly and a new and final highest-level set of needs arise: the pursuit of intellectual and aesthetic gratifications. Some of those who made it to the top of this needs hierarchy became intrigued by the intellectual notions of ecology and the

40

aesthetics of nature. Their attitudes and political leanings shifted to new ecology-oriented lifestyles

This "needs hierarchy," as detailed in Maslow's *Motivation and Personality*, holds a scientific explanation of why economic arguments have lost much of their persuasive power. More and more people have risen to the intellectual and aesthetic levels of the scale, while industry arguments have not risen correspondingly, but remain anchored for the most part in basic economics. People inhabiting the higher levels tend to either be unaware of or to scorn arguments from what they perceive as lower levels, and such people are now sufficiently numerous to constitute a political force. More than a few of them are firmly entrenched in the federal environmental bureaucracy.

POST-GRATIFICATION PATHOLOGY

Most importantly, Maslow discovered that rising all the way up the needs hierarchy has consequences. People at the highest levels tend to feel an "independence of and a certain disdain for the old satisfiers and goal objects," says Maslow, "with a new dependence on satisfiers and goal objects that hitherto had been overlooked, not wanted, or only casually wanted." Old gratifiers "become boring, or even repulsive." New ungratified needs are overestimated. Lower basic needs already gratified are underestimated or even devalued. The old camera or stereo isn't good enough any more. The old consumerist lifestyle isn't good enough any more. The old worldview of economic necessity isn't good enough any more. We must have a New Environmental Paradigm. In one crucial passage Maslow solved the mindstyle of the New Environmental Paradigm:

> In a word, we tend to take for granted the blessings we already have, especially if we don't have to work or struggle for them. The food, the security, the love, the admiration, the freedom that have always been there, that have never been lacking or yearned for tends not only to be unnoticed but also even to be devalued or mocked or destroyed. This phenomenon of failing to count one's blessings is, of course, not realistic and can therefore be considered to be a form of pathology.

41

Maslow called this condition "post-gratification forgetting and devaluation." He said prophetically of it, "This relatively neglected phenomenon of postgratification forgetting and devaluation is, in my opinion, of very great potential importance and power." How right he was. That phenomenon is the source and head of the environmental movement. Maslow never lived to see America in the age of environmentalism. It would be interesting to know what he would think of our present oversupply of pathological fools who devalue and mock and destroy our resource base, our industrial might, our fate as a nation.

ENVIRONMENTALISM AS RELIGION

With this basic anatomy and nurture of environmentalism in hand, we are ready to probe the dynamics of environmental behavior to find a frame of reference that will make sense of its peculiarities. For purposes of explanation, I have divided environmentalism into two major parts, wildernism (an ideology encompassing the conservation movement, the nature appreciation trend, anti-technology and so forth), and regulationism (a statist political style popularized by many kinds of socialists and also by environmentalists such as John Muir, founder of the Sierra Club, and built into the U.S. federal government during the New Deal era). In reality, the two are inextricably meshed, but it is helpful to separate them for clarity of thought.

Wildernism can best be understood as a religion. Many environmentalists who are not members of any religion object to this assertion, but their behavior belies their words, and Milbrath's analysis of environmentalism as a soteriology reinforces my conclusion. Years of personal intuitive experience and more years of careful objective research have convinced me that this idea is correct: wildernism behaves like a religion.

I was first put onto this track by wildernist literature itself. Book reviews in wildernist journals such as *The Living Wilderness* and the *Sierra Club Bulletin* contain scores of titles such as *A Theology of Earth* by Rene Dubos, *Ecology and Theology* by Gabriel Fackre, and *The Spiritual Aspects of Wilderness* by Sigurd Olson. Phrases such as "reverence for life," and sentiments such as "we shall not be able to solve the ecological crisis until we recapture some kind of spiritual relationship between man and his environment"are common.

Scientific studies of wildernism, for example the Forest Service's *Wilderness Users in the Pacific Northwest—Their Characteristics, Values and Management Preferences*, use a "wildernism attitude scale" containing a measurement significantly labeled "esthetic-religious." In short, there is compelling evidence upon which to ground an investigation of the religion-like behavior of wildernism.

That leads us to ask: what, then, are the usual characteristics of a religion so that we may see if wildernism possesses all or most of them? A concept of deity is common to most but not all major religions—neither Buddhism nor Taoism, for example, envision a Supreme Being. But wildernism does embrace a concept of deity, one similar to pantheism. Pantheism is a doctrine that the universe (nature) conceived of as a whole is God, a view that runs through most environmentalist writing, even the earliest, as any reader of Thoreau's *Walden, or Life in the Woods* will recognize. Belief in pantheism is not mandatory, but belief in the sanctity of earth and its life is.

Wildernism, like many religions, claims the powers of salvation and healing, asserting that "In wildness is the preservation of the world," and that the active wildernist enjoys better physical health, gains a sense of spiritual renewal, obtains relief from the emotional and psychic tensions of "dirty civilization" (the "psychic safety valve" theory of wildernism), and gains a general peace of mind available even to non-using believers who benefit from just knowing wild areas are safe somewhere in the world.

Wildernism offers a sense of community (Maslow's need for a "sense of belonging") through group rituals such as day hikes, extended outings, pack trips, rivers runs and so forth, as well as in regular group meetings, rallies and demonstrations back in civilization. This reinforces the "us" and "them" feelings that are helpful in fighting material progress and "saving" wilderness.

Wildernists run an evangelical missionary service to the "Great Unwashed" through their sophisticated and literate publications programs. Correctly gauging the power of the better educated and more intellectually and aesthetically oriented population, wildernist publications employ writers of literary excellence, use poetry to reinforce gorgeous nature photography, have a deep sense of history, a biting humor, bitter cynicism, and don't mind bending the truth or prostituting science if it helps the cause. Their evangelizing also extends to lobbying their morality into law.

Wildernism has a hierarchical priesthood. I don't mean just the elect few who have been lionized by the media such as "Arch-Druid" David Brower; I mean the zealous minions on the paid staffs of environmental organizations including the executive directors and lobbyists (mostly attorneys), and the hundreds of volunteer board members, chapter, group and committee chairpeople who, by my experience, are selected as much for their ideological devotion as for their administrative ability and personal ambition.

To a wildernist, as to many of the religious, any compromise or backing off on substantive issues amounts to a moral stain. Lying for the cause, which goes against most religious systems, is common among environmentalists as merely a means justified by the ends, a sentiment that has disgusted me many times in environmentalist committee meetings.

Wildernism, to summarize, provides most if not all the characteristics of a standard religion as recognized by scholars of the subject: a sense of distinctiveness and community, of awe and cosmic unity, standards of morality and irreproachable beliefs, rituals, tests of faith and grounds for expulsion, a central dogma that must be protected, and so on. The adherents of wildernism are convinced of their moral and ethical superiority, are blind to reason on questions of dogma, and feel that they have an exclusive hold on the truth. It all adds up to religious behavior, and one does not expect objective rationality from religious behavior, one expects devotion and, at the extreme, zealotry.

This explains why a wilderness controversy is such a profoundly personal thing to a wildernist, a phenomenon I have heard industry workers wonder at. Industry's perception that such fights are recreation-versus-jobs, or symbolically, work-versus-play, is ridiculously superficial. Industry strategy based upon that perception uniformly loses in court and in Congress. Wildernist propaganda in dozens of issues from the

Monongahela to the Redwoods has used this misapprehension to make industry look like a mob of crass and insensitive boors with no concern for the public interest, the inner life, or higher values.

THE ULTIMATE ROOTS OF ENVIRONMENTALISM

This descriptive analysis of wildernism defines the problem better and gives us a firmer grip on environmentalist motives, but it does not account for the origins of environmentalism or the basic mechanism that makes it tick. Where did it come from? How did it start? Why does it persist?

These questions have fascinated me for years. As far back as I look in the history of civilization I find some evidence of regulationism (simply because governments appeared when civilization arose), anti-civilization, and other "environmentalist" sentiments, although the elements of wildernism, such as appreciation for strictly wild places as opposed to mere "nature," are missing until about 700 years ago. Despite extensive writings about "nature" as a force or a god, nowhere in the ancient history of Egypt, Mesopotamia, Israel, Greece or Rome do we find the slightest love for wilderness, or even wild scenery for that matter. In Rome, we even find bitter complaints that the earth contains wild places at all: Lucretius (100 B.C. - 55 B.C.) railed in *On the Nature of the Universe* that part of the earth "has been greedily seized by mountains and the woodland haunts of wild beasts. Part is usurped by crags and desolate bogs." Worse, "the little that is left of cultivable soil, if the force of nature had its way, would be choked with briars, did not the force of man oppose it."

Poets beginning with the Greek Theocritus (316 B.C. - 246 B.C.) did praise **pastoral** landscapes. These landscapes were not urban, not wild, but a cultivated rural "middle landscape," as modern historian Leo Marx calls it. Not until the writings of Petrarch, father of the Renaissance, near the end of the Middle Ages (1336 A.D.) do we find in literature an actual love of wild, uncultivated places. Something like aesthetic appreciation of wild landscapes can be found in Oriental literature many centuries earlier, but with heavy nature-worship overtones.

Why should some elements of environmentalism be so

recent while others are so ancient, as old as civilization? Is our analysis wrong? The key to this question, in the view of historian Roderick Nash, may lie millions of years earlier than civilization. Noting the scientific findings of anthropologists such as L.S.B. Leakey, Nash points out that as the primeval forests of central Africa receded in response to climatic changes roughly fifteen million years ago, our prehuman ancestors abandoned the trees to dwell in the grasslands. In this new habitat, says Nash, humans evidently evolved the upright posture which enabled us to peer over the grass, the sharp eyesight and depth perception to compensate for the superior speed, size and sense of smell of our competitors and predators, and the large brain to anticipate, plan ahead and survive. The sense of sight in this new and dangerous habitat became crucially important, Nash tells us.

The point of this evolutionary excursion—and my friends who dismiss evolution as a flawed theory, pay attention anyway, for Nash's conclusion is interesting regardless what the truth may be—is this: if grassland vistas were the cradle of humanity, then *forests and other visually limited places were the original wilderness.* Visual limitation meant danger and death. Millions of years of dependence upon visual acuity, says Nash, have stamped a lasting bias on the mind of humanity.

Nash's conclusion: "Appreciation of wilderness must be seen as recent, revolutionary, and incomplete. Friends of wilderness should remember that in terms of the entire history of man's relationship to nature, they are riding the crest of a very, very recent wave. Ambivalence, a blend of attraction and repulsion, is still the most accurate way to characterize the present feeling toward wilderness."

Undoubtedly, this bias against wilderness, whether evolutionary or not, shows up very early in documentable history: it probably helped spur the agricultural revolution that made settled life possible in western Asia about 11,000 years ago. In addition to this prejudice against wilderness, there are ancient writings from the dawn of civilization that show prejudice against material progress and civilization itself. One marvels at the modern tone of the cry in the Mesopotamian *Epic of Gilgamesh* from 2,000 B.C.: "Here in the city man dies oppressed at heart. I have looked over the wall and I see the bodies floating on the river, and that will be my lot also. . . . I will go to the country where the cedar is cut." Cities were already places of discontent in this, the earliest piece of coherent literature on earth, considerably predating the Bible.

PRIMITIVISM

Arthur O. Lovejoy clarified the questions raised by the *Epic of Gilgamesh* in his *Primitivism and Related Ideas in Antiquity* by identifying the perverse streak of regressiveness known as "cultural primitivism." This primitivism is "the discontent of the civilized with civilization, the belief of men living in a relatively complex cultural condition that a life far simpler is a more desirable life." Lovejoy also pointedly notes that "nature" is the most slippery term in human language: he proves it by elaborating 66 different definitions of "nature" as found in ancient literature.

Lewis S. Feuer notes in his *Ideology and the Ideologists* that "primitivism, an enduring theme of human nature ever since the civilizing process began, probably began, as Lovejoy conjectured, when early men sought out the refinements of cave-dwellings, and brought down upon themselves the first primitivist protest . . . Now it is a universal law that every ideology tends towards a revival of primitivism."

Primitivism, the age-old regression away from conscious responsible management of our lives as civilized human beings, is as close to the rock-bottom source of environmentalism as I can discover. Primitivism versus progress has been a main theme of all human history.

Today we find primitivism hidden in such slogans as the motto of the Sierra Club: "Not blindly against progress, but against blind progress." It sounds noble at first because we expect that if one is only against *blind* progress, and is also *not* blindly against progress, then one must be *for* sighted progress. That, alas, does not follow either in logic or in the Sierra Club, which does not grant the existence of sighted progress. No matter by what elegant sophistry it is disguised, the primitivist core of that motto is still the word **against**. Negativism and obstructionism are of the essence of environmentalist ideology.

Now we have begun the barest exploration of environmentalism. We know its necessary opposition to free enterprise somewhat more completely and realistically. The complex anatomy of environmentalism tells us not to underestimate its sophistication; its behavior like a religion tells us to set it apart from "just folks" thinking; its roots in primitivism tell us that it is no passing fad, but a perennial problem. Lest we continue to dismiss emotionalism's power, we should heed social psychologist Gustave LeBon's observation:

"Reason creates science; sentiments and creeds shape history."

3

Taking the Offensive

WHAT CAN AMERICAN INDUSTRY DO about the harmful effects of environmentalism? In practical terms, there are only three rings in the environmental circus where the outcome of issues can be influenced: the arenas of (1) public opinion, (2) legislative and administrative lobbying and pressure, and (3) litigation—the courts. Sophisticated, committed environmentalists have taken the initiative in all three combat zones and have made sweeping gains during the last two decades, while American industry has mostly taken the defensive. The fate of defensive teams is well known.

I have noticed as a consultant to the private sector that when an industry is attacked by an environmentalist organization, high level managers seem to lose their minds. They say such things as: "We won't stand for this. Let's go on the offensive. Let's tell *our* side of the story!" Those last two sentences have nothing to do with each other. If you fail to see why, consider what happened when the United States was attacked by Japan at Pearl Harbor. Did President Roosevelt say: "We won't stand for this. Let's go on the offensive. Let's tell *America's* side of the story"? No. The president in effect said: "Kill the bastards!" (with no offense intended to our contemporary Japanese friends). My point is this: The conflict between environmentalists and industry amounts to an undeclared war, and industry will never win by merely telling its side of the story.

As an observer with experience on both sides of the conflict, I think American industry needs to fight fire with fire—to gain more sophistication, commitment and initiative. This is not a call for mere press agentry and public

49

relations flackery making broadside attacks on environmentalism. That would fail. The public cannot yet tell the difference between the environment and environmentalists. The day will come. Remember, it's not too hard to grasp that ecology is not environmentalism.

But for now, I'm making a recommendation for a tough intellectual effort on the part of American industry to understand a good bit more about itself in today's socioeconomic picture than most businesses do, and then to follow it up with mobilization. A good start was made during the Reagan administration's early years with the deregulation effort which was partially successful and the privatization effort which went just about nowhere. The key fact that I'm trying to get across is that an increasing number of the major forces affecting business today are social and political, not just technological and market forces.

INDUSTRY IS TOO TIMID

The proper approach to handling these new pressures has been a matter of intense industry controversy, ranging from the hard-hitting advocacy campaign of Mobil Oil Corporation whose vice-president Herbert Schmertz is "not out to make people love us. We want to stimulate and participate in public dialogue, to show our deep intellectual commitment to the subject matter;" to the moderate activist campaign that Bendix Corporation carried on during the tenure of William M. Agee who made himself available to the public and felt that part of his job was to be "a public figure and to take positions on public issues, not just company activities;" to the ultraconservative view that the bottom line tells it all, as held by Gene M. Woodfin, chairman of Marathon Manufacturing Co., who says, "a lot of drum beating most likely hides a company that's not as sound as it would like to appear."

The bulk of American industry sentiment tends closer to the views of Marathon than Mobil, and I think that could lead to serious long-term difficulties. As Louis Banks of M.I.T.'s Sloan School of Management has noted, no longer is simply doing our jobs—producing products, making money, and owing the world nothing else in terms of justification—a sufficient rationale for dealing with powerful social and political forces such as environmentalism. This does not mean that industry must bow to opponents or adopt some socialist approaches to a private welfare state. It means aggressively attacking those opponents with intelligence and skill to show that the private sector, the profit motive, private property and the rest of the free enterprise catechism are what built America. Many citizens who are innocent bystanders to ecology wars are not aware that the free enterprise system built America and allowed her citizens to prosper as none before.

New York consultant Osgood Nichols puts his finger on the main point: "The power, respect, and freedom to act of an institution are based on whether or not it is working, and is perceived to be working, in the interest of the citizenry." To put it quite bluntly, an ignorant public made irate by anti-capitalist assertions that we are evil profit-mad monsters could put us all out of business.

THE POWER OF IDEOLOGY

Our basic resource industries—farming, livestock, mining, fisheries, forestry, petroleum exploration and extraction, among others—perhaps more than others are faced with an intellectual challenge. Environmental ideologies, those big ideas such as the "land ethic" and the "ecological conscience," have become pervasive cultural symbols that quietly feed public support for indiscriminate wilderness withdrawals and other measures that will damage American industrial productivity. Since the problem of ideology is vastly more rarified than the typical engineering or economic issue, many in industry tend to pooh-pooh its importance. Failure to appreciate the power of ideology, however, could be disastrous.

Environmental ideology tends to show up in unexpected but critical areas, such as new theories of jurisprudence that favor nature over humanity in the law. This insidious type of anti-humanism is best exemplified by U.S. Supreme Court Justice Douglas' dissenting opinion in the case of *Sierra Club v. Morton* (1972), in which the Sierra Club sued to revoke the U.S. Forest Service permit granted to Walt

51

Disney Enterprises, Inc., for construction of a $35 million recreation complex in Mineral King Valley adjoining California's Sequoia National Park.

"Contemporary public concern for protecting nature's ecological equilibrium," wrote Justice Douglas, "should lead to the conferral of standing upon environmental objects to sue for their own preservation. This suit should therefore be more properly labelled as *Mineral King v. Morton*."

You don't have to be a lawyer to figure out the economic chaos that would result if every tree in a logging operation and every stone in a mineshaft could sue you as Justice Douglas would have it (endorsed by Justices Blackmun and Brennan). Luckily, the Sierra Club lost its suit, which could have granted devastating powers to environmental groups, but only by a technicality and only by *one vote* from the bench. As it was, the decision did grant environmentalists the right, once they had established their direct stake in such a suit, to claim the public interest in defense of natural values, a decision which seriously weakens recognition of the public interests served by industry. Neither the forest nor livestock industry did so much as to file an *amicus curiae* (friend of the court) brief in this case to support human values, either because they did not know of Disney's difficulties or did not realize that the outcome could have affected more than the entertainment sector. The final disposition of Mineral King Valley was settled only in 1978 when President Carter signed the $1.2 billion Omnibus "parks barrel" legislation adding the valley to Sequoia National Park.

Industry executives and workers will beef about specific issues, but during the 1970s, they simply didn't *want* to know about environmental ideology, an attitude which came back to haunt them a few years into the Environmental Eighties. Years ago, these industry leaders wanted to think of their opponents as kooks and weirdos, as birdwatchers and little old ladies in tennis shoes. Since Congress and the courts have kicked that notion out of them, they have overshot the mark just as far in the opposite direction and now want to believe that environmentalists are merely honest but misguided "plain folks."

Environmentalists are nothing of the sort. How many

losses in Congress and the courts will it take to prove that environmentalists form a hierarchy of clear-eyed, pragmatic hired executives and lobbyists carrying out the will of a religiously inspired primitivist volunteer leadership which in turn guides a mixed bag of zealots, interested supporters, misguided plain folks and faddish hangers-on *who are winning with ideology?*

COMMITMENT

I have seen at first hand the commitment that has replaced the rah-rah fervor of the early '70s in environmentalist ranks. As social scientist Amitai Etzioni discovered, "In our attempt to explain why some societal actors realize more of their goals than others—even though these others command similar knowledge and assets and act in basically the same environment—the higher commitment of the actors who realize more of their goals seems to be a major factor."

I don't see that commitment in American industrial ranks, primarily because any industry is not a monolithic whole. Take the forest industry as an example: it is "an industry" in name only. In fact, it is a ragtag assortment of more than 10,000 diverse and highly competitive business enterprises constrained by myriads of governmental restrictions, not the least of which is a clutch of anti-trust laws. Further, serious internal problems fragment this industry: regional jealousies, self-serving cliques, small-operator alienation from large corporate viewpoints, labor friction, non-activist leadership, and lack of communication—all these problems serve to shrivel the ringing phrase "the forest industry" into a pale abstraction.

A substantial part of the problem is industry's tendency to buy off environmental activists through foundation grants. The non-profit charitable foundations of great private sector firms have provided more than half the total income of environmental groups since the early 1970s. Yet the financial support these same foundations have given to grass roots pro-free enterprise groups is piddling or absent. You don't have to be very smart to see what lies in the future for private sector liberties when the Arco Foundation, the Rockefeller Brothers Foundations, the Weyerhaeuser Foundation, and dozens of others fund the enemies but not the advocates of profit, private property, and individual freedom.

The private sector's ability to think or act in unison is virtually non-existent. On the occasions when solidarity has surfaced, as when the inter-industry coalition formed to fight the Alaska D-2 lands bill in the late 1970s, I've stood up and cheered. Numerous industry critics point out that if it were not for a handful of powerful trade associations, American industry would have no defense against environmentalists at all. It takes a big sexy issue to get even a tiny fraction of American industry's leaders together, and the record shows that when the big issues die down, industry promptly goes back to public affairs sleep.

TRYING TO COPE

The economy's leaders, of course, are not unaware of these problems. The historical solution is for each industry to band together in trade associations, an idea which was pioneered by the forest industry, and which works quite well as long as the mainstream culture is not too activist. Forest-related associations provided the model for other industry associations. The first national trade association in the United States was the Writing Paper Manufacturers Assn., founded in 1861 at Pittfield, Massachusetts. Then came the American Paper and Pulp Assn. in 1878, followed by dozens of others. A recent directory of forest-related associations alone listed more than 125 separate entities, but the real powerhouses are the "Big Five:" the National Forest Products Assn. (1902); the American Forest Institute (1931); the American Plywood Assn. (1933); the American Pulpwood Assn. (1934) and the American Paper Institute (14 old-line trade associations merged to form API in 1966).

Top elected and staff officials of the "Big Five" in turn make up part of the Forest Industries Council (1943), a committee that analyses problems and charts strategies, although its actions are not binding on its member institutions. Also, the forest industry once a year brings a hundred or so ranking executives together for a few days to form the Economic Council of the Forest Products Industry, which is as close to an overall coordinating committee as any industry has. Again, no decisions taken by this council are binding on any member firms or associations. Many other industries such as petroleum and mining bring together top executives to form similar loose councils once or more each year. It is amusing to note that Soviet analysts believe that "monopoly capital," as they like to call the American system, must be guided by some hidden central control point, they are so blinded by their own totalitarian system of Marxist-Leninist central planning. They would be utterly appalled at the actual degree of disorder and disarray in American industrial leadership. They would never be able to admit that such a leaderless mess of corporations is precisely the source of American capitalism's economic vigor, a vigor that perpetually eludes Karl Marx's deadlines for its collapse. These trade associations and councils are the primary reason that industry is able to count as many victories as it can in the combat theaters of public opinion, lobbying and litigation.

Public affairs is covered for the forest industry by its educational arm, the American Forest Institute. For years AFI has run one of the most effective and respected public affairs programs in American business, but as we shall see, that may be faint praise. AFI's programs are intelligently planned and implemented. Campaigns are not dreamed up according to the pinches felt by any one firm or sector, but rather result from testing the public waters to see what the issues really are as best demographers can tell, and the results are sometimes surprising, as they were in late 1978.

Using surveys by such prestigious pollsters as Opinion Research, Inc. and Yankelovich, Skelly and White, AFI discovered that 73 percent of the public believed most paper and lumber firms harvest trees in a responsible way, and that the forest industry was surpassed in public fondness only by agriculture for doing a "good job" of resource management. These cheery findings came side by side with the revelation that most Americans (1) don't have the foggiest idea what officially designated wilderness really is, especially in regard to limitations on access and recreation, and (2) aren't remotely concerned about lumber and paper shortages.

This left the industry in the bizarre position of facing a rather friendly public with no capacity whatever to accept a general anti-wilderness stand justified by potential product shortages. So how do you fight an idiotic smile condoning massive land grabs that could cripple or destroy an industry?

KEEPING A LOW PROFILE

AFI decided to play it cool, to not alienate the public's existing good feelings toward the industry, "to strengthen and enlarge the favorable public climate in which the industry operates," as AFI vice-president Jim Plumb told me at the time. Most forest products firms active in public affairs followed the same plan, which in the early 1980s proved to be disastrous as more and more wilderness designations were made and more and more valuable timberlands were removed from commercial management.

AFI could not see the disaster coming, though. Plumb assured me that AFI would not duck the hot potato issues such as intensifying forest management on federal lands or the controversy over the phenoxy herbicides. However, AFI's handling of such issues in fact led to *lessening* of intensive management. Their gentlemanly program was shoved aside by the hot emotionality of environmentalist charges that intensive management creates irreparable damage, as exemplified in restrictions added to timber harvest plans on the Six Rivers National Forest in California. AFI's velvet gloves were torn to shreds in a storm of intensive management-related lawsuits in the early 1980s brought by the Sierra Club, the National Wildlife Federation and others, as in the salmon-habitat damage case that halted all timber harvest on the Mapleton Ranger District of Oregon's Siuslaw National Forest in 1984.

Then, too, when the facts began to show that anti-herbicide sentiment had as an important component the protection of commercial marijuana crops on "guerilla plantations" in National Forests, AFI avoided the subject like the plague and proved totally ineffectual in stopping the Environmental Protection Agency's emergency cancellation ban on 2,4,5-T.

AFI ads still stick to low-key, soft-sell messages accenting the positive, carefully targeted to the 14.5 million Americans already known to be political activists who read the usual newsweeklies like *Time, Newsweek,* and *U.S. News and World Report.* AFI also taps the "opinion leader" audience in such journals as *Smithsonian, Saturday Review, Harpers,* and the *New Yorker.*

AFI's ad campaigns are only one of five basic elements in their total program. They also operate educational programs in classrooms across the nation, the American Tree Farm System to encourage better forest productivity on private individually owned lands, a publications program including the well-known poster-sized newsletter *Green America,* and finally, a sophisticated press relations operation with key members of the media which tries to obtain fair and comprehensive coverage of industry views on hot issues. I have seen numerous positive results of this media relations operation in clipping service editorials from prominent newspapers.

Despite the high marks AFI's program gets with large timber corporations and the public relations industry, there is a sizeable population of critics, particularly among small loggers who do not belong to the "Big Five" associations, and who mistrust AFI's soft-spoken approach as being too namby-pamby, lacking the grit and gristle to handle tough issues decisively.

For example, one well-placed industry source informed me that when the Forest Service released its RARE II recommendations favoring 36 million non-Wilderness acres out of a 62 million acre total study area, and the environmental lobby instantly went public alleging a "sweetheart deal" with the forest industry, AFI held off its response (thereby giving credibility to environmentalist arguments) because the hard numbers to support its case for larger non-Wilderness acreages hadn't come out of the computer yet. My source accused AFI of being too timid to speak out on the common sense issues without statistics to hide behind, despite the fact that public input had overwhelmingly favored a much larger non-Wilderness area.

REACTIVE OR PRO-ACTIVE?

A number of loggers and even middle managers in large timber corporations feel the need for a more aggressive advocacy campaign like that of Mobil Oil. Mobil has spoken out in ads on public policy, energy deregulation, profits, and has even made tart replies to media errors, naming names and detailing the nitty gritty. Mobil has also sponsored the milder "Observations" column in selected newspaper Sunday supplements, which usually deals with some ideological issue in a manner designed to de-fuse fanatic environmental beliefs. A recent column reviewed a book by a previously devout environmentalist who had second thoughts, and now wants to set people straight on the facts. If my poll of public affairs departments in the top 10 timber firms means anything, such a scrappy intellectual issues-oriented campaign horrifies the over-conservative woods business.

Some fascinating proprietary information comparing the relative effectiveness of Mobil's advocacy campaign with the AFI soft-sell approach was leaked in 1978 and published in a leading business review. It gives the conservatives a pat on the back. Mobil's ads, so the report tells us, achieve 90 percent penetration, that is, a creditable 90 percent of the intended audience actually read their ads, but only 33 percent agreed with them. AFI, on the other hand, the review goes on, had much poorer penetration, but 56 percent of the government leaders and 70 percent of the public polled said the AFI ads were useful in supplying them with information of forestry issues. The report quoted Mobil critics as saying "You don't want 90 percent of the people reading ads that only 33 percent agree with. You can't win a vote with 33 percent."

However, neither the review nor the survey it refers to noted that while the majority of AFI's readers said the ads were useful, no comment was made about whether readers agreed with the useful information, or which side they would use it on. Actually, the comparison was completely meaningless, apples and oranges, Mobil's "agreement" rating vs. AFI's "usefulness" rating.

Is the soft-sell better than Mobil's abrasive advocacy? Maybe not. One survey tells us that the forest industry ranks second only to banking as a socially responsible industry. The catch is, the same survey found that no industry got more than a one-third vote of public confidence. Another survey revealed that the forest industry had improved its standing in public regard as having done a "good job" conserving natural resources for several years in a row. The catch is, the best year, 1978, found the timber industry had only reached a 34 percent "good job" rating, while about three out of ten answered the same question by saying the industry had done a "poor job" conserving resources.

Isn't the timber industry being damned with faint praise by these demographics? Aren't we fooling ourselves by thinking these results are "good?" Is the soft-sell's 34 percent any better than Mobil's 33 percent, especially considering Mobil's powerful, even antagonistic arguments? Isn't there some venturesome timber firm out there willing to risk being a public gadfly for a time and really explore the issues in contemporary social dialogue rather than politely seeking agreement? I think not.

INTELLIGENT ATTACK

The payoff of an aggressive issues-oriented campaign isn't immediate, says M.I.T.'s Louis Banks, but is "the soundest long-range strategy for business-media relationships," and the effort involved may be the decisive factor in assuring the freedom of large corporations to act as independent centers of management decision making in the future. Whether U.S. business likes it or not, says Banks, it "lives in a capitalist democracy, and the virtue of the *capitalist* factor must continually be redefined and argued in current social context if the public *political* factor is to overcome its populist suspicions of corporate power and scope."

Outside academia, past chairman Reginald H. Jones of General Electric Co. backs up Banks' views on making "externalities" a regular part of total management approaches: "Chief executive officers of all major corporations will have to become activists rather than adaptive. There will be no room for Neanderthals."

Executives in the forest industry would do well to take this advice to heart. Many have. For example, Jay Gruenfeld, while he was vice president of Potlatch Corp., pioneered environmentalist-industry discussion groups to solve problems with communication where feasible. Leaders of

Union Camp Corp. and Simpson Timber Co., among others, work closely with The Nature Conservancy to make selected lands available for preservation. Georgia-Pacific provides quarterly videotape programs called "In Focus" to its employees on everything from environmental problems to benefit programs, with enthusiastic reception. St. Regis Paper Co. distributes annual calendars with pertinent themes such as "the tree in art," a sensitive exposition of aesthetic awareness that has drawn favorable note from a wide audience. Weyerhaeuser and Arcata, among others, support the long-term educational programs of the Forest History Society, which helps set the record straight through rigorous scholarship. Every company has its own management style, but all could benefit by leaning more to public interest activities.

The associations of the "Big Five" do an incredible job in the lobbying arena, but even here gaps arise. I was appalled to find that not a single soul in either of the two biggest timber lobbying groups, NFPA and API, had the slightest idea how many forest industry bills won or lost compared with how many environmentalist bills won or lost in Congress for any given year. It would not take a great deal of reshuffling to keep tabs on the 300 to 400 environmental bills that are plunked into the Congressional hopper each year, and this scorecard would give everyone a much better idea of the real results from all the marvelously researched programs managed by AFI.

The forest industry also needs to keep track of current U.S. Court of Appeals and U.S. Supreme Court suits that might produce decisions unfavorable to the industry, regardless of who the litigants were. Remember, it was Walt Disney Enterprises, Inc. that got into the tussle with the Sierra Club which nearly lost a decision that could have destroyed the American forest economy by giving human rights to natural objects.

And what about the small logger? Many are reluctant to speak out in defense of their industry. I think they are victims of stereotyping, and mistrust their wildernist counterparts who are educated middle class urbanites who enjoy characterizing loggers as irresponsible and inarticulate clods.

Hogwash. Long experience has shown that it is the plain, unvarnished truth spoken by plain, unvarnished citizens, not statements by full-time lobbyists or public communicators, that is most persuasive in shaping public opinion. The most sophisticated strategy the industry could employ is to boost these rural loggers into the public arena.

YOU GOT THE TREES, WE GOT THE VOTES

And if you rural loggers think the industry doesn't need your help, let me remind you of California state senator Bill Green of Los Angeles' Watts district, who warned: "Your activities may be rural, but your problems are urban. You have the trees, but we have the votes. Your problem is what my constituents *think* you are doing in the woods, accurate or inaccurate." The industry's finest ought to let them know how it really is, so, loggers, get off your axe and tell it your way. The least you could do is set aside a few dozen bucks a year for Mailgrams to public officials on selected issues. And large corporations and associations, how about sending interested loggers your backgrounders on the important issues so they'll have the benefit of topflight analysis?

Outside the forest industry there are groups ready to form coalitions with forest-related interests, big ones such as the American Farm Bureau Federation and the National Assn. of Home Builders. There are smaller ones such as the National Council for Environmental Balance, a group of qualified scientists who are trying to undo the propaganda of their fellow scientists who prostitute truth for environmental ideology and personal gain. NCEB is prepared to provide outstanding speakers for conferences, to exchange information, and to assist in defending a viable economy in a livable environment, according to president Dr. Irwin W. Tucker.

There are dozens of good ideas such as these waiting to be enacted, but the problem is not thinking up good ideas, it's putting them into effect. That is entirely a matter of **commitment**. The environmentalists have a great deal of it. I see precious little of it in the forest industry, and what there is must grow stronger.

And never have I seen an industry with the guts to stand up and attack environmentalism with the intelligence and conviction it will take to overcome the statist thrust. Many are the Senators and Representatives who had told me, "Arnold, your friends in industry just don't have any guts. They come out with all these positive messages about how important industry is to the American way of life and never tell us how awful the environmentalist ideal really is. We're always looking over our shoulders to the left, but never to the right. Does industry think we're fools in Congress? Or are they feeble-minded?" Well, my industry friends, which is it? Our defeat be on you if you don't wise up.

Somehow, I think you'll wise up. Eventually. In the meantime, it is important that everyone in American industry know the social and political basics, things like the value of history, how environmental propaganda works, and how to counter it. It is important that everyone in American industry know how to be a political activist, regardless how humble or exalted, and to have a clear idea of all the positive good free enterprise is doing in the face of environmentalist deception and public suspicion. This all adds up to gaining sophistication, to developing commitment, and to taking the offensive.

4

History:
The Unconventional Weapon

Fellow citizens, we cannot escape history.
—— *Abraham Lincoln*

WHETHER IT'S SOME OBSCURE LAW such as the Antiquities
Act of 1906 being invoked to give President Carter authority
to set aside Alaska lands without Congressional vote or the
2nd Circuit Court's 1965 decision in the case of *Scenic Hudson Preservation Conference v. Federal Power Commission*
that gave environmental groups their standing to sue in defense of scenic, historical, and recreational values, it is true
that as Honest Abe said, we cannot escape history.

American industry managers by my experience are more
inclined to want to make history than to read it. While this
bent has led them to technical and economic power, it has
done little to equip them for dealing with the social and
political power—and growing economic power—of environmentalism. The shape of the past, as environmentalists well
know, is crucial to the shaping of the future. Without an
historical memory, or what a lawyer would call a body of
precedent, American industry will lack an overall strategic
sense and a feel for the political relationships necessary to
combat rampant ideological environmentalism.

Using history as a weapon in ecology wars may seem
plebian and disgusting to purists, but note that environmental lawyers have no qualms about wielding a sharp historical rebuttal. Industry would be foolish not to use every

63

equalizer it can get, especially in the face of an increasingly well-educated and informed public.

IN SEARCH OF HISTORY

A few years ago I got an impressive lesson in the scope of industrial history and the impact it can have on modern society. Retrace with me a 25,000-mile American odyssey as I produced a film for the Forest History Society—and was disabused of the notion that history is a dry scholarly pursuit of no particular significance or use in the "real world."

And don't think I'm about to dish up some old Marxist historical materialism, which insists that history is a force of nature that follows scientific laws and can be predicted. History does not contain that kind of meaning. Marxists believe that determinist, dialectical, materialist forces exist within the universe which *inevitably* push society through six progressive "socio-economic formations" beginning with primitive communism (tribal societies), then slave-based societies, then feudalism, then capitalism, then socialism, and finally, someday in the future, "full communism." This theory, called "historical materialism," asserts that human consciousness *inevitably* progresses higher with these "socio-economic formations" and that human control over nature *inevitably* progresses with them, too.

It's worth examining this classic internal inconsistency within Marxist philosophy. If the universe is totally materialistic and there is no God, as Marx repeatedly insisted, then how can matter recognize the superiority of one "socio-economic formation" over another? How can matter be so cooperative that it selects "the working-class struggle" as the basis for social evolution? How can matter arrange social history in such a neat pattern that communism inevitably wins and capitalism inevitably loses? Obviously, in a totally materialistic universe it can't. Idealism, the notion that ideas have force over the universe, a bugaboo which Marx denounced again and again, sneaked into Marx's own thought in "historical materialism." As Neil McInnes pointed out in his "Marxist Philosophy" entry in *The Encyclopedia of Philosophy*, Marx never resolved this dualism:

64

After a profession of materialist faith orthodox Marxism introduces the idealist element by attributing to matter a readiness to cooperate with progressive causes. (In other contexts such an attribution of spiritual purposes to matter is called magic.)

I think that history is not a progressive movement of consciousness or of the control by man over nature as Marx said. The kind of history I found suggests that **economies** follow the rule of utility and that **politics** follows the rule of legitimacy and that **culture** follows the rule of Maslow's Needs Hierarchy. That leads to specialization and structural differentiation, if it leads anywhere at all, and there are no simple, determinate relations between the three realms. The course of history is susceptible to human intervention because history as I use it is nothing more than the record of human intervention.

WHO'S GOT HISTORY?

And now to my historical wanderings: Forest History Society Executive Director Dr. Harold K. "Pete" Steen called me to his headquarters office in Santa Cruz on California's southern redwood coast to discuss the society's project. The Forest History Society has subsequently relocated its headquarters to Durham, North Carolina, but its mission remains the same. The society promotes a wide range of historical activities from preserving important records to publishing works by qualified historians on the industrial, environmental and governmental aspects of the forest, refereed by a panel of scholars to maintain rigorous standards of accuracy and objectivity.

"People are skeptical about history," said Steen. "How can we show its true value?"

I pondered, recalling the sentiments voiced by Carl Sandburg in his well-known poem, *The People, Yes:*

> "Bring me my liar," said a king
>   calling for the historian of the realm.
> "History is bunk," said a history-
>   making motor car king.

65

"History is a fable agreed upon,"
said a shriveled smiling Frenchman.
"Even if you prove it, who cares?"
demanded an Illinois state librarian.

Taking my cue from Sandburg, I wondered who does care about forest history. "Pete," I said, "give me your membership list and let me talk to the writers and users of forest history. That's your story."

Thus began my long journey criss-crossing North America in search of forest history. At the University of Maine at Orono, history professor David C. Smith gave me the first long perspective of his field. "History is thought by some people to be a game played on the past, but it's not that at all. History is really used by everyone every day to make the decisions they face about how to deal with their future, how to manage their money, how to manage their companies, and how to develop policy. For example, here in Maine we outlawed log runs on our rivers to avoid environmental damage. Now we're discovering that the large network of roads required to truck those logs out has a worse impact on our soils than log runs had on our rivers. The lesson we learned is history at work."

You'd expect an historian to pat history on the back. But what about the leaders of industry, environmental groups and government? What do they think about history?

In Federal Way, Washington, George Weyerhaeuser, president of the timber company that bears his family name, said about history: "It is absolutely essential. We're in a business that requires long-term perspective. We acquire and manage lands as a perpetual resource base, we plant trees requiring 25 to 50 year life cycles, and we run manufacturing complexes near our lands as permanent infrastructure. A knowledge of history is vital as it provides the necessary background and sense of continuity from which our plans for the future evolve."

In San Francisco, Sierra Club Executive Director Michael McCloskey told me, "No field is changing as fast as the environmental field. To be able to see where we're going, we need to know where we've come from, what our past looks like, and what its meaning is. As we look ahead to the 1990s, we need to be historians of contemporary history to understand what's happening and what we face in the years ahead." (He was still speaking to me at the time. He's too good for my sort of free enterprise activist trash these days.)

In the Russell Senate office building in Washington, D.C., Senator Mark Hatfield showed me his collection of Lincoln memorabilia, and recalled the President's epigram, "We cannot escape history." The Oregon senator sat and mused, "Lincoln had a great capacity to lift the familiar to the realm of the profound. But I'm not certain that we always heed his words. I say that because so frequently in the legislative process, we move on purely contemporary understanding. If we had that historical perspective and used it, we would develop far better legislation. For example, I recall the debate on a bill concerning clearcutting. I reminded my colleagues that long before the white man came, the Indians used clearcutting by fire to enhance the growing area for their food plants. Meanwhile the wildfire created the forest openings so necessary to the regeneration of the Douglas fir, which needs mineral soil and direct sunlight to reproduce properly. There is wisdom to be learned from history."

A DASH OF POLITICS

And a great deal of strategic understanding as well, I discovered. In the Mobile, Alabama offices of International Paper, Fred Gragg, public affairs officer at that time, said, "The forest industry, particularly here in the South, very early provided leadership and capital to establish a second growth forest. But the public is almost completely unaware of the tremendous pioneering efforts of industry, and accept the claim that industry's only role was to cut out and get out."

Gragg showed me an issue of the society's publication, the *Journal of Forest History*. An article detailed the life of Henry E. Hardtner, owner of Urania Lumber company and Louisiana's first conservationist. As early as 1909, Hardtner was replanting his harvested lands and warning his fellow lumbermen to "protect your remaining forests and commence at once the reforestation of your denuded areas."

Gragg concluded, "Because of their scholarly credibility, these articles can go a long way toward correcting the public's misunderstanding about the industry's role in conservation in America."

While the idea might horrify scholars, history is a goldmine of intellectual ammunition in the fight for free enterprise against centralizers, statists and socialists of all kinds. It is an ideal tool to handle the criticism that industry is too rationalistic, aloof, and unemotional, symptoms that arouse suspicion and kill credibility. History can soften the alienating impact of formulas and graphs without softening factuality. Forest history is not merely the record of who did what in the forest when, it is a living time line of the big ideas, attitudes, and beliefs that captured the hearts and minds of our forefathers and shaped their lives, and therefore ours.

FORWARD TO THE PAST

In North Carolina's Pisgah National Forest, I took a morning stroll into conservation history at the "Cradle of Forestry" exhibit of the U.S. Forest Service. Here Carl Alwin Schenck, German-born and trained forester, founded the Biltmore Forest School in the fall of 1898, the first forestry school in America, on what was then the George W. Vanderbilt estate. While managing Vanderbilt's lands, Schenck schooled his "Biltmore Boys," as he fondly called them, in the then unorthodox principles of profitable sustained yield forestry.

As I walked through the rhododendron-studded forest, I could almost hear the echoes of long-vanished hoots and hollers in the restored bunkhouse, cook shack, and classroom. The trail I walked had been trod not only by Schenck, but also by his predecessor at the Vanderbilt estate, the man who coined the term "conservation" and later went on to become the first chief of the U.S. Forest Service in 1905: Gifford Pinchot. That fact reminded me of the quirks of history that can color a whole generation's outlook on environmental matters.

Historian Samuel P. Hays researched Pinchot's era for the origins of the conservation movement. The result, *Conserva-*

tion and the Gospel of Efficiency, disclosed the view that our usual idea of conservation as a wave of popular sentiment for wilderness preservation and wildlife protection is misleading. Hays claims that conservation actually developed in the 1890s as a scientific movement led by a small group of professional men (including Pinchot) whose objective was the orderly, efficient use of resources under the guidance of experts, a principle they called wise use. By this definition, I see the forest industry as the world's foremost conservationist.

Later, however, during the 1908 battle between Pinchot and Sierra Club founder John Muir over a reservoir to be built in California's wild Hetch Hetchy Valley, the term conservation was gradually expropriated by advocates who felt that current conservation projects should abandon the wise use theory for the perpetual preservation theory. As time went on, the vociferous preservationists became identified with the word conservation, and this is the sense in which a wide public understands the term today. The historian thus reminds us that we must look to the sources for correct understanding.

HISTORY AS ANALYSIS

But historians do not limit themselves to setting the record straight—they interpret it, evaluate it, and make it answer questions such as why the conservation movement occurred at all, and what its consequences were. These questions have a direct bearing on understanding the explosion of environmentalism in the 1960s and '70s.

The seeds of conservation were planted in the massive industrialization of America after the Civil War ended in 1865. Smoke from factories, coal-heated homes, and locomotives became an everyday fact. Rivers bore increasing loads of sewage, and mining and timber cutting were more and more in evidence. The prophecies of George Perkins Marsh's 1864 book *Man and Nature* were coming true: the inexhaustibility of the earth was a myth. In short, it had become impossible to ignore man's impact on the environment. The responses were many.

Landscape architect Frederick Law Olmsted became an

early advocate of urban parks—he designed New York's famous Central Park and the U.S. Capitol grounds—as a means of resisting the "vital exhaustion" and "nervous irritation" of the city, in which, Olmsted said, "We grow more and more artificial every day." In 1865 he proposed that patches of "wild forest" be preserved near metropolitan centers.

Then came the conservation groups. America's first national conservation group was founded in 1848 by John Wesley Powell, hero of the first daring river run through the Grand Canyon during a hydrographic reconnaissance—the American Association for the Advancement of Science, which took early note of water supply, forest preservation, and land use issues. The AAAS was followed in 1870 by the American Fisheries Society, in 1875 by the American Forestry Association, in 1888 by sportsman Teddy Roosevelt's Boone and Crockett Club, in 1892 by the Sierra Club, and in 1900 by two groups promoting beautification and preservation: the American League for Civic Improvement and the Society for the Preservation of Historical and Scenic Spots.

Urban parks and membership groups were not the only evidence of growing concern for the environment—legislation also advanced. In 1872, Yellowstone became the first wilderness-like National Park "for the benefit and enjoyment of the people." (Although Hot Springs National Park had been created in 1832, it was originally a military reserve and never became much more than a spa for the elite. While Yosemite was granted to California in 1864, it was not made into a state park until 1890 and didn't become a national park until 1905.)

In 1875, Congress passed the first law protecting a wildlife species, the buffalo. By an ironic twist of fate, President Grant vetoed it because the buffalo hunters were better at starving the Plains Indians than the Army was at killing them in combat. The first national forest legislation was enacted in 1891, authorizing the president to create forest reserves on federal lands to conserve timber and water and to prevent floods.

But it was an historian who first linked the wilderness with sacred American virtues. Frederick Jackson Turner in an 1896 *Atlantic Monthly* article asserted that the American, "Out of his wilderness experience . . . fashioned a formula for social regeneration—the freedom of the individual." Thus wilderness was seen as a positive force fostering independence, confidence in the common man, and self-government. In 1903, Turner noted that the "rough conquest of the wilderness is accomplished." But if an urban-industrial civilization was replacing the wilderness, where would our strength and purpose come from?

Modern historian Roderick Nash said in his *Wilderness and the American Mind* that it was with "a considerable sense of shock" that Americans realized what Turner was saying. What had seemed the forward march of civilization now appeared as urban sprawl. Where the wilderness had brought forth sobriety, honesty and hard work, the city now seemed to spawn decadence, immorality and squalor. A tone of pessimism and nostalgic regret crept over America.

In one of the classic overreactions of all time, America rushed to form a cult of the wild. The Boy Scout movement of Sir Robert S.S. Baden-Powell was founded to retain the invigorating influence of wilderness in modern civilization. Pacifists worried that the surplus of national energy that had conquered the frontier might be turned to conquering the world. Nature writers like John Burroughs, John Muir, and Jack London drew huge followings. Teddy Roosevelt's second State of the Union address asserted that "forest and water problems are perhaps the most vital internal questions of the United States." *Boston Post* headlines sensationalized the 60-day-stunt of part-time illustrator Joe Knowles, who one August day in 1913 stripped naked and trudged into the Maine woods to live off the land "as Adam had" (he succeeded). Other headlines repeated the idea that we would run out of everything within a decade or two, especially forests.

REPEATING HISTORY

If our great grandfathers had known their history a little better, they would have been neither so surprised at the close of the frontier nor so concerned about its outcome. Urbanization, resource depletion, and substitution are age-old themes in classical literature. Consider these sentiments: "My boss will have his mullet, imported from Corsica or from The rocks below Taormina: home waters are all fished out. To fill such ravening maws, our local breeding-grounds, Are trawled without cease, the market never lets up." An Audubon Society lobbyist? No, remarks from *The Sixteen Satires* by Juvenal, Roman poet, circa 100 A.D.

Or consider this: "The rich, soft soil has all run away leaving the land nothing but skin and bones. For some mountains which today will only support bees produced not so long ago trees which when cut provided roof beams for huge buildings whose roofs are still standing." John Muir bewailing redwood logging? No, it is Plato of 3rd century B.C. Greece in his dialogue *Critias*.

But although the people of antiquity may have admired nature, they never loved wilderness. Plato talked of "nature" not as wild, uncultivated landscapes—nor as developed urban scenes—but rather as what modern scholar Leo Marx called the "middle landscape," not urban, not wild, but a rural, cultivated scene so familiar in bucolic poetry, which originated with Theocritus about 300 B.C.

The Judaeo-Christian tradition was obsessed with wilderness as the seat of evil. Remember, Moses and his contrary people were condemned to wander 40 years in the wilderness, and Jesus went into the wilderness to be tempted by Satan. They equated wilderness with desert, and were so obsessed by its hardships that the ancient terms for the evil wilderness occur 245 times in the Old Testament and 35 times in the New Testament.

Other ancients were more worried about productive land, like the Roman philosopher Lucretius, who wrote in about 60 B.C., "of all that is covered by the wide sweep of sky, much has been greedily seized by mountains and the woodland homes of wild beasts. The little that is left of cultivable soil, if the force of nature had its way, would be choked with briars, did not the force of man oppose it."

Longinus, actually an anonymous First Century A.D. Greek author given the name by a scribe's error, wrote a critical treatise *On the Sublime* describing the immensity of nature, of the stars, of the mountains and volcanoes, and of the ocean, as a source of the sublime. As pointed out in *The Oxford Companion to English Literature* (Fifth Edition), a 1674 French translation influenced many British writers such as Addison, Hume and Burke and prompted the "cult of the sublime," a growing aesthetic appreciation of the grandeur and violence of nature. The cult of the sublime encouraged enthusiasm for wild scenery and cosmic grandeur in the Eighteenth Century. It eventually led away from the rationalism of the Enlightenment and toward Nineteenth Century Romanticism, with its emphasis on feeling and imagination. But we're getting ahead of ourselves.

The fact is, nobody in all of Western civilization said a kind word about wilderness until Petrarch in 1336 A.D. Perhaps it took that long for man to feel secure enough in his civilization to perceive wilderness as aesthetically gratifying. By the time Sir Walter Raleigh's explorers landed in Virginia in 1584, Arcadian nature poetry had become popular enough to influence Captain Arthur Barlowe into seeing the New World as paradise regained.

"We put ashore," he wrote, "and found great abundance and plentie. The woods are full of the highest and reddest Cedars of the world. The natives be more gentle, loving and faithfull, voide of all guile and treason, and such as live after the maner of the Golden Age." This has to be regarded as a sort of Elizabethan travel hype to attract colonists, because when William Bradford stood off from Plymouth in the *Mayflower* in 1620, he saw "a hidious and desolate wilderness, full of wild beasts and willd men." But then, Bradford had to stay there.

COLONIAL ENVIRONMENTALISM

As early as 1626 conservation laws appeared when Plymouth Colony forbade sale of timber out of the colony without approval of the governor and council. In 1674, the Massachusetts Bay Colony passed environmental legislation to forbid pollution of Boston harbor. In 1681, William Penn insisted that for every five acres of forest cleared, one acre should be kept in trees. And in 1710, the Massachusetts Bay Colony forbade the creation of "any disturbance or encumbrance on or across any river that would operate to stop or obstruct the natural passage of fish."

In 1729, re-enacting a 1691 policy, the White Pine Act put the forest at center stage in act one of the drama of the American Revolution. The "King's Broad Arrow" was branded on the best trees in the colonies, reserving them for the Royal Navy's masts and spars. The colonists, of course, cut the marked trees first, paying no attention whatever to the new conservation law, preferring the king's trees, since their quality had been certified by an expert.

About this time, English attitudes about wilderness slowly began to shift away from the idea that it was the home of Satan where the faithful might revert to savage and bestial ways. John Ray's 1691 *The Wisdom of God Manifested in the Works of the Creation* even suggested that mountains might be the handiwork of God, if not his very image. However, the practical American settler's attitude toward wilderness was to transform it into the rural, middle landscape, not wild, not urban. Most pioneers took seriously the instruction of Genesis 1:28 to "Be fruitful and multiply, and replenish the earth and subdue it," to have dominion over every living thing.

STURM UND DRANG: THE ROMANTIC ERA

The Enlightenment of the 18th century brought belief in the inevitability of progress and confidence in mankind's advance over ignorance and the uncivilized. It later shaped the scientific utilization theory of conservation. But a 19th century movement, Romanticism, erupted in Germany, France and England as a reaction against the critical analysis and rationalism of the Enlightenment, and later fostered the perpetual preservation theory.

Only five years after Lewis and Clark finished their epoch-making 1806 exploration of the Northwest, William Cullen Bryant wrote the first American nature poem, *Thanatopsis*, finding moral and religious significance in "the continuous woods where rolls the Oregon." Beginning about 1825, the Hudson River School of painters, Thomas Doughty, Thomas Cole, Albert Bierstadt, Asher B. Durand, and half a dozen others including telegraph-inventor Samuel F. B. Morse, forwarded this Romantic theme in wilderness landscapes. Washington Irving and James Fenimore Cooper introduced the European Romantic movement to American literature with such stories as Cooper's *Leatherstocking Tales* and *Light in the Forest*. In his 1825 *A Forest Hymn*, Bryant wrote "the groves were God's first temples," foreshadowing Ralph Waldo Emerson's 1836 essay *Nature*, which introduced the philosophy of Transcendentalism, a belief that "nature is the symbol of the spirit," that is, the real world is basically spiritual.

Soon the frog of Walden Pond began to croak the same tune. Henry David Thoreau said in a lecture in 1851, "In wildness is the preservation of the world." By 1854, he had completed his hymn to living a wilderness life in *Walden, or Life in the Woods*, in which he rhapsodized on his frugality in spending only twenty-eight dollars and twelve-and-a-half cents for materials to build his cabin and 27 cents a week for food.

When I visited Walden, I was disillusioned. It lies scarcely a mile from the town of Concord. A local resident, who is also a professor at M.I.T., informed me that Thoreau had walked the pleasant stroll to his mother's kitchen in Concord several times a week for a free dinner, which he didn't count in his 27 cents. The author even admitted it in his book. In the chapter entitled *Economy*, Thoreau lamely writes, "To meet the objections of some inveterate cavillers, I may as well state that if I dined out occasionally, as I always had done, and I trust I shall have opportunities to do again, it was frequently to the detriment of my domestic arrangements." (He never explained what domestic arrangements at Walden could have been harmed by frequent free dinners, since he lived alone and essentially had no domestic arrangements.)

Even John Muir, who deeply admired Thoreau's philosophy, could not suppress a chuckle upon visiting Walden to find it "a mere saunter from Concord," and smiled at a man who "could see forests in orchards and patches of huckleberry brush." But Thoreau forever set the tone of inspired poetic exaggeration that blossomed in the turn of the century's conservation/cult-of-the-wild movement, and thrives today in environmentalist circles. And I believe it was Keats who said "Poets are the unacknowledged legislators of the world."

THE PROFESSIONALIZATION OF CONSERVATION

Within a few years the influence of Thoreau and other nature lovers entered the professional world through such men as Frederick Law Olmsted, and the die was cast. The wilderness movement became an American institution. Membership groups soon grew like toadstools after the rain. The Appalachian Mountain Club (1876), the Sierra Club (1892), the Mazamas of Portland, Oregon (1894), the Save the Redwoods League (1922), and the Wilderness Society (1935). Although the conservation movement lost some popularity by 1920, the administration of Franklin D. Roosevelt revitalized the ghost of the wise-use idea in statist programs such as the Civilian Conservation Corps and the Tennessee Valley Authority. Socialistic though they were, these government enterprises nevertheless reflected concern for expert management of natural resources as Pinchot had originally envisioned it.

History gives us the ability to understand that the growth of environmentalism was a long, slow process, to see the patterns in its growth, such as the replay in the 1960s of Romanticism displacing Rationalism, as happened at the beginning of the 19th century, and to see the historical legitimacy of the forest manager's own wise-use practices.

CAN INDUSTRY FIND HISTORY?

The forest industry is gradually learning to use history to its own benefit. In New York, communications executive James Kussmann of St. Regis Paper Company before it was taken over by Champion International, described the use of the firm's own history to enhance new employee orientation. In Neenah, Wisconsin, Menasha Corporation executive Mowry Smith, who wrote his firm's history, told me how it had become a valuable reference tool in planning for the future. In Boca Raton, Florida, Louis A. Huber of Lumbermen's Underwriting Alliance said that his firm looks for an historical sense in mill owners seeking insurance, as it indicates the urge to continuity and affects safety, health and fire prevention behavior.

At San Diego State University, history professor Thomas R. Cox capped off the role of the Forest History Society in changing the American mood. "The attitude in university circles toward business several decades ago tended to be very hostile. In those days, writings on forest history normally dealt with land frauds, robber barons, and all the nasty things businessmen had done. Today, there's been a decided change in tone. One now approaches timber firms, land policy, and other things without all of one's political allegiances sworn in advance—at least a good scholar does. I think the Forest History Society has to be given credit for this change, probably more than any other single agency."

But don't expect the Society to take sides. Its objectivity and insistence on factuality is the source of its value to the combatants. As Society Executive Director Steen says, "The historian's analysis of failures along with successes at times brings charges of sensationalism or, even worse, irreverence." I've seen these charges come from industry, environmentalists, and government in about equal amounts. But remember, the job of the society is to find out what *really* happened, not to pad the truth for any interest.

Environmentalists are gaining skill in cost-benefit analysis, resource policy, and how to harass business, all tricks they learned from the historical record. All of American industry must do no less in combating them. If knowledge is power, history is a weapon in gaining power, albeit an unconventional one. Modern philosopher George Santayana reminded us of a thought voiced by the Greek general Thucydides 2,400 years ago:

Those who are ignorant of history are doomed to repeat it.

5

The Truth About
Resource Management

THE TRUTH ABOUT RESOURCE MANAGEMENT is this: it is no
longer simply a technical and economic pursuit. Today it is
social and political as well, replete with institutionalized citizen activism, an entrenched environmental bureaucracy, and
a coterie of lobbyists who would like to see industry hogtied,
it not butchered. The result for resource managers has been
headaches that grow into migraines and skull fractures.

THE PRIMACY OF VALUES

Value judgments have entered the picture in an important
way. Where we used to ask "**How** can we increase resource
productivity?" voices now inquire "**Should** we increase resource productivity?" Large ethical questions have also become hot political weapons. Congressional floor debates, law
texts, and Supreme Court decisions increasingly reflect suspicions of man's technological impulses and the propriety of
man's domination of the earth.

Technology and economics seem increasingly irrelevant to
advocates. In fact, an essay by George Hall, *Strategy and
Organization in Public Land Policy*, asserts: "Indeed, at times
the argument seems to be that only aesthetic and moral considerations are relevant."

The legitimacy of industrial civilization is being seriously
eroded by the powerful onslaught of primitivist environmental ethics, values, beliefs, and attitudes. In the process,
the resource manager's motives are getting a black eye and
the knot of government regulation is ever tightening around

79

his freedom to manage, in part because these ethical and moral questions have never been answered. It is now time to deal with the ethics of resource management.

If ethics seems too remote and abstract to affect the outcome of issues, let me share with you an environmentalist secret that hits a key resource industry blind spot: it is values, beliefs, and attitudes (the components of an ethical system), not facts and information, that rule in social and political processes. Donald Scherer emphasized this point in *Personal Values and Environmental Issues*: "The conclusions people reach about what they ought to do are not based simply on an examination of scientific or advertisers' "Facts." Such decisions regularly reflect values." Yet our resource industry has an almost obsessive reliance upon facts alone.

Many years ago when I sat in the councils of environmentalism, we were keenly aware of this. In the Wilderness Act controversy, we stressed the "land ethic" (a set of values) and the industry opposition whipped out a chart showing board feet locked up (a set of facts). In the North Cascades National Park dispute, we pushed the "ecological conscience" (a set of values) and the industry opposition responded with statistics about job losses (a set of facts). In the 1968 Redwood National Park fight we hit aesthetics and the industry opposition came out in a newspaper ad pitting "Sierra Club Claims" (a set of values) versus, what else, "Industry Facts." This is not to belittle the importance of factuality and truthfulness in winning social conflicts. It is to demonstrate the industry's unawareness that practical politics today must address more than facts—it must address values, too.

Values are the basic defensible units of any social struggle. If you don't believe it, read Judge William E. Doyle's decision in the 1970 case of *Parker v. United States*, or Judge Maxwell in the 1973 case, *West Virginia Division of Izaak Walton League, Inc. v. Butz*, both of which stopped timber sales from the National Forests—the decisions are rank with values. Using a value-based strategy stressing what **ought to be** rather than what **is**, environmentalists fired the public's imagination and consistently defeated resource industries during the 1960s and '70s despite objections to "emotionality."

80

How Values Operate

Why did it work? Scientific studies show that the answer lies in certain basics of human nature. A value, as psychological researcher Karl E. Schiebe points out in *Values and Beliefs*, is a statement of what is good, what ought to be, what is preferable. A belief is a statement of what is possible, what exists, what happened in history and the like. Attitudes are simply packages of beliefs. But there is a crucial difference in the power of values and the power of beliefs and attitudes. Social scientist Milton Rokeach put his finger on it in the study *Beliefs, Attitudes and Values*: "A value, unlike an attitude, is an imperative to action, not only a belief about the preferable, but also a preference for the preferable." In short, values have motivational power. Beliefs and attitudes do not. Beliefs (concerning what is) tend to fall in line behind values (concerning what ought to be).

Facts and information were found to be effective in changing beliefs, and attitudes. But when it comes to changing values—the real sources of social and political action—only ideals, philosophy, religion, emotions, and similar forces work. Unless you provide some dramatic gut level appeal, you will not attract social or political support. As communications sociologist Hugh Dalziel Duncan wrote in *Symbols In Society*, "People do not want information about, but identification with, community life. In drama [real life struggles between good and bad principles of social order] they **participate**." According to this analysis, resource industries made a social and political blunder when they neglected drama, objected to emotionality as a tool of its own, shunned the rationale of values, and raised rationality to irrational levels.

Environmentalists Shift Gears

Some observers have noted that environmentalists are now arguing their cases not on the emotional concerns but on the scientific merits. According to a 1979 *Business Week* article, Arlie Schardt of the Environmental Defense Fund says it is "doing more cost-benefit analysis to justify the environmental programs it proposes." During the early and mid-1980s, environmental groups perfected their cost-benefit tactic. As of this writing, in many cases groups are demanding economic justification for management activities slated for federal lands, having realized that Forest Service and Bureau of Land Management policy has never been based on free market economics but more on Keynesian welfare economics and bureaucratic empire-building.

Does this indicate a sudden love for the free market, corporate capitalism and individual liberty on the part of environmentalists? Not at all. Environmental leaders in particular are still statists of the worst rank. It's simply a new tactic to cope with recessionary trends and the conservative swing that came with the two Reagan administrations.

Environmentalists feel they have the liberty to migrate into economic policy because their values have been well rooted in modern society for years: a Resources for the Future poll in 1979 found that 53 percent of those questioned believed that "protecting the environment is so important that the requirements and standards cannot be too high." The new economic policy analysis tactic is merely a buttress shoring up the primitivist fortress of the environmental ethic.

SOME INTROSPECTION

Then what is industry's answer? First, we must make the effort to reflect on industry's own values, beliefs and attitudes, and to realize that in an ethical sense it is as worthy as any human activity, and then to offer its own ethic in society's marketplace of ideas along with all the factual and technical justifications it uses now. But this is easier said than done, as I discovered when I first became acquainted with working resource people, particularly foresters, loggers, and managers 15 years ago as a conservationist delegate to forest "show-me" tours. Our discussions tended to be very practical, site-specific, and technical. Basic human values never came up, even seemed alien to the industry mentality, while among environmentalists such topics were commonplace.

The cultural differences between environmentalists and industry members, I found, were much more than simple class differences. It was not the old Marxist saw about the proletariat (working class) struggling against the bourgeoisie (capitalists). That struggle, such as it was, went on between industry workers and industry managers. The environmental battle is something completely different, not class warfare but cultural warfare. Environmentalists fight for what they regard as "big ideas." Most industry workers and industry managers, on the other hand, *are not idea oriented*, and have little contact with "big ideas" either through discussions or reading. Many loggers wouldn't be caught dead in a library. Many forest managers have never come in contact with the notion that there might be **moral and philosophical principles** behind their way of life.

But after some years of getting close to forest industry folk, learning who they really were, and tipping a few with them at the local pub, I began to realize that these industry people were *living* their values, not talking about them. True, the education of the average logger is only nine years (half that of surveyed environmentalists), so some had neither the analytical training to identify and classify their basic values nor the verbal polish to articulate them concisely. But what came shining through was their native intelligence, their deep feeling for the land, for their work, for their friends, for their communities. These were good people, something that many environmentalists would be loath to admit except in the most condescending tone.

Yet getting at their values would still occupy me for nearly a decade, going behind the scenes, asking questions, quietly observing. And now I'm ready to hold up a mirror to these people who are my friends to show them—and the world—values that can change history, values that may well provide us with the guidance to survive the turbulent decades of conflict that lie ahead.

CREDIT WHERE IT'S DUE

Conventional wisdom has it that the forest industry—that all industry—is reactionary and the enlightened environmentalists are going to drag it kicking and screaming into the 21st century. I found early in the game that things are the other way around: industry is doing the acting, and environmentalists are doing the reacting. Environmentalism is essentially anti-progressive and ultra-reactionary, but masquerades in the most popular words it can find. Consider: in the woods I found ingenious logging systems gathering raw materials for civilization's basic needs, food, clothing, and shelter (yes, dozens of foodstuffs contain wood derivatives as emulsifiers, stabilizers, sweeteners, and more, while acetate and other fabrics begin life in the forest). These logging systems ranged all the way from horses to helicopters, thoughtfully adapting technology to dozens of variables, such as soil types, climate, time of year, landforms, timber types and sizes, local economics, equipment availability, regulations, and environmental constraints. An essential element of everything I saw was **creativity**, the pure basic urge to bring something into being, to be competent in the physical world.

In Idaho's white pine forests I saw horses yarding logs on sites too sensitive for motorized equipment (the horses even had to wear diapers to prevent water pollution!). In North Carolina's piedmont woods, I saw high flotation skidders plugging through swampy woods and over dikes to bring the goods to truck landings. In Montana I watched a feller-buncher grab 22-inch diameter trees in its mechanical hands and snip them off at ground level with a giant shear. In Mississippi, I witnessed whole trees go into a total tree chipper and come out the other end as perfectly shaped and sized wood chips for pulp and papermaking. In Maine's spruce forests, I saw low-ground-pressure log skidders treading lightly on delicate terrain—and in Oregon, I even saw such a machine "flown" across a canyon on a logging skyline so that a serious blowdown on the far side could be retrieved for lumber without having to build environmentally disruptive logging roads on a public watershed. And in California I gawked at a chopper lifting a turn of logs clear of the ground without so much as a scrape and flying it to the truck landing. All across America, I saw people struggling to be more productive and more protective.

The forester and research scientist were innovating, too. With the shrinking land base of commercial forests, which is being nibbled away by Wilderness designations and urban sprawl, they have been applying hard-won knowledge to developing techniques that will grow more wood faster and healthier to meet increasing demand. The corporate manager uses computer modeling to predict the optimum directions for the huge capital investments required to regenerate the forests of the future, investments that will show no return for decades, and then only at terrible risk of loss from fire, insects, disease, and other disaster. The mill operator is right in there with them, striving to get the last measure of usable product from available raw materials with thin-kerf saws to make narrower cuts and leave more wood, and finding that he can also become energy self-sufficient by conversion of the stored sunlight in wood wastes to electricity by steam generation. Truly, what I saw was man's creative and rational urge doing battle with the gods to reconcile the needs of civilization with the quality of the natural environment—a worthy goal in any ethical system.

THOU SLUGGARD

But in environmentalist circles, I found no such ingenuity, nor approval, nor even suggestions for improvement, only peevish complaints against each one of these technical solutions to a basic problem of civilization. It is plain that the environmentalists are the true reactionaries, ultra-reactionaries, when it comes to thinking about civilization. As Claus and Bolander note in *Ecological Sanity*, "All too many of the leading environmentalists are themselves filled with inertia and lack imagination," and can only cry for "bans" and "de-development." But these same dullish faultfinders can be inventive, subtle, and worldly-wise when it comes to politics. There is an ideological reason for their obstructionist stance: anti-technology and anti-civilization values.

For all the good and valuable aspects of environmentalism, there is a dark side that must be critically examined if we are not to act as fools. Anti-technology is the most obvious facet of that dark side. Anti-technology abhors the "technological fix" to environmental problems. Technology itself, you see, is evil. Rene Dubos put the dogma this way in *So Human An Animal*: "Modern man is anxious, even during peace and in the midst of economic affluence, because the technological

85

world that constitutes his immediate environment, by separating him from the natural world under which he evolved, fails to satisfy certain of his unchanging needs." Dubos incorrectly blames technology for what is actually a natural part of human motivation, the urge to rise up the "needs hierarchy" as described by Abraham Maslow. As one human need is satisfied, another higher level ungratified need takes its place—and brings on the "post-gratification forgetting and devaluation" we discussed in Chapter One. That is the real source of generalized anxiety. We can find cases of "weltschmerz' ("world-pain") and "existential dread," to use the fancy jargon of pedants, in non-industrial primitive cultures. They experience the same anxiety "during peace and in the midst of economic affluence" as some of us do.

The point is, no matter how good the technical solution, it will not do, because environmentalists do not want to solve technical problems, they want to eliminate technology. So the forester's genetic enhancement is condemned for improving growth and productivity; the logger is berated for suspending his logs on skylines to protect creekbeds; the mill operator is degraded for utilizing former waste scraps for new products. It's just more of the same technological fix to an environmentalist.

But the truth is, there isn't another kind of fix: any plan humans implement is some kind of technology, even the "behavioral fix" of government directives so beloved of environmentalists, even leaving nature alone—which conveniently doesn't fix civilization's problem, it abandons it. And so we see that devout environmentalists are not interested in making civilization work better, even though they ungratefully benefit from it: their diatribes are printed on paper made from trees by industrial processes; their carping photographs of "sacrilegious" clearcuts are taken on film made from trees by industrial processes. They want a return to a simpler, non-industrial existence that does not resemble civilization. The fact that returning to non-scientific agriculture and non-industrial techniques would doom much of the world's population to starvation, disease and death doesn't seem to matter.

WARTS OF THE WORLD

Anti-civilization is another hidden ideology of environmentalism. This belief, which many environmentalists hold only half-consciously, harks back to the mythical Golden Age described in classical Greek and Roman literature, which was envisioned as a heroic time of perfect harmony at the beginning of the world when nature provided all human wants—everything afterward was only a falling away from this ideal. Clarence Glacken traced the history of this strong yearning for mythical simpler times, for nature itself to provide all necessities without human sweat or toil, in his magnificent book, *Traces on the Rhodian Shore.*

But the child-like desire for a Golden Age world did not die with the Roman Empire's writers. It re-emerged in the Renaissance in Italian pastoral poetry and can be found in Elizabethan writings such as Sir Philip Sidney's *Arcadia.* It can even be found in Shakespeare: *The Tempest* contains a marvelous scene in Act II in which the talky lord Gonzalo conjectures how, given the opportunity, he would rule the desert island upon which he and his fellows have just been shipwrecked:

> I' the commonwealth I would by contraries
> Execute all things, for no kind of traffic
> Would I admit, no name of magistrate.
> Letters should not be known; riches, poverty,
> And use of service, none; contract, succession,
> Bourn, bound of land, tilth, vineyard, none;
> No use of metal, corn, or wine, or oil;
> No occupation—all men idle, all;
> And women, too, but innocent and pure;
> No sovereignty —
>
> SEBASTIAN: Yet he would be King on't.
>
> ANTONIO: The latter end of his commonwealth forgets the beginning.
>
> GONZALO: All things in common nature should produce
> Without sweat or endeavor. Treason, felony,

87

Sword, pike, knife, gun, or need of any engine
Would I not have. But Nature should bring forth
Of it own kind, all foison, all abundance,
To feed my innocent people. . .
I would with such perfection govern, sir,
To excel the Golden Age.

These thoroughly anti-civilization feelings were wide-spread in Western Europe during the late 1500s; Shake-speare lifted much of this scene from one of Montaigne's essays, *On Cannibals*. And how like an environmentalist the scene sounds: the Sierra Club certainly wishes to admit no traffic anywhere (except its own), does not feel the "need of any engine," and seems to believe that no work is necessary to live—they destroy so many jobs. And when it comes to politics, they'd obviously like to "be King on't," to rule us all. And the latter end of the environmentalist common-wealth, a total regulation state, forgets its beginnings in simple love of nature, much as the good but foolish Lord Gonzalo.

This anti-civilization trend re-appeared in the Romanti-cism of the mid-19th century, in the conservation movement of the turn of the 20th century, and blossomed yet again in the environmental movement of the 1960s. Today we see its stamp in bans and regulations that restrict the means of pro-duction, yet take no thought of what this will do to the yield, as if somehow nature would compensate for the economic loss, or as if you could have your cake and eat it too. The anti-civilization argument is seldom fully thought out or blatantly stated, yet remains an insidious element of en-vironmentalism.

NAIVETÉ AND GUILE

Maslow's "needs hierarchy" tells us that the disquiet Rene Dubos found in modern civilization comes more from un-realistic expectations than from dynamic progress, and more from seeking after ineffable fulfillment than from the crea-tive force of industrial technology. The fact that environmen-talists come almost exclusively from the service sector and not from the working class also has something to do with their anxiety about civilization and their need to cleanse their psyches in wild nature: they're so far removed from daily physical labor or the grind of real work they've become hypersensitive. As Shakespeare said through the voice of Hamlet, "The hand of little employment hath the daintier sense."

Although the rank and file environmentalist sincerely embraces anti-technology and anti-civilization as personal values, these beliefs are only a veneer hiding the true motivations of their organizational leaders. As Gonzalo would be King on his untrafficked Golden Age isle, so do ambitious environmental group leaders seek unending political clout for their groups and want totalitarian power for their ideas. Remember that many such leaders are lawyers who have been taught in law school to go for the jugular and have great skill at lobbying and litigating their viewpoints into ever more restrictive laws. Unfortunately for the public, no single piece of their agenda may appear to point toward totalitarian power, but when you assess the cumulative effect of any fifty environmental laws, as our media never do, the trend is unmistakable—and intolerable.

The aggressive leadership of the environmental movement adds to the totalitarian problematic. This leadership began with dedicated amateurs who took home less pay than many waitresses, people such as David Brower in his youth. But more and more we are seeing professionals, attorneys and politicians at the helm of environmental groups, people getting upwards of $100,000 a year such as the National Wildlife Federation's Jay Hair.

And what do these people believe about cherished American values such as private property, corporate capitalism and individual liberty? One can get an idea from what they studied in law school. The casebook of Hanks, Tarlock and Hanks, *Environmental Law and Policy*, spends its first 90 pages parading forth essays that challenge these basic American values: Garrett Hardin, *The Tragedy of the Commons*, which asserts that industrial capitalism engenders ecological destruction and that free enterprise must therefore be replaced with "mutual coercion mutually agreed upon," and Lynn White, *The Historical Roots of Our Ecological Crisis*, which preaches that the Judaeo-Christian religious tradition with its Biblical precept of man's dominion over all other creatures has contributed to man's devastation of the earth (one is tempted to ask Dr. White how China and India got the Jews and Christians to come and deforest the Yellow River Valley and the drainages of the Indus and Ganges with their religious beliefs). The philosophical underpinnings of the modern environmental ethic are fundamentally opposed to free enterprise and strongly biased toward centralized government control.

WEEVILS AND SHIPBORERS

During the Carter administration, many environmental group officials found employment with the federal government, people such as Barbara Blum, Richard Cotton, Leo M. Eisel, Katherine Fletcher, Angus MacBeth, J. Gustave Speth and Larry Williams. A regular "ecoligarchy" grew up to plague Washington—one Washington observer, Llewellyn King, called it the "termite infestation." Most of the big termites were swept out with the Reagan election, but many lower level bureaucrats still give their political allegiance to the Sierra Club or the Wilderness Society before the flag of the United States.

There was great hope among some during the Carter administration that a genuine political takeover might some day be possible. Ernest Callenbach's two novels *Ecotopia* and *Ecotopia Emerging* described an environmental dictatorship in the Pacific Northwest replete with all sorts of Marxist-Leninist features. I might also remind you of Marshall McLuhan's claim that writers, poets, and other artists form a DEW line—they give us a "Distant Early Warning" of what is happening to our culture.

If it seems reasonable to ask, "Won't these environmentalists calm down when they realize how much they've already won?", let me ask you "Won't you stay sated once you realize how much you've eaten?" The environmental movement is institutionalized in enormous multi-million dollar non-profit corporations that need a constant stream of money to survive and that means a constant stream of issues, of "environmental threats" in order to raise that money. Don't believe it? Let me quote you a fascinating passage from Lewis Coser's *The Functions of Social Conflict*:

> Struggle groups may actually 'attract' enemies in order to help maintain and increase group cohesion. Continued conflict being a condition of survival for struggle groups, they must continually provoke it. . . so the groups' search for enemies is aimed not at obtaining results for its members, but merely at maintaining its own structure as a going concern.

My personal experience verified that this "morality" of the expedient is rampant in environmental organizations.

DEAF, DUMB, AND BLIND

The final straw that snapped my former association with the Sierra Club came one night at a conservation committee meeting where this crass policy reared its ugly head. A member had on a hike found a setting of fell-and-bucked logs dislodged by a storm and lying in an Oregon creek. The logs came from the property of a large and well known timber firm. This member introduced a motion to submit his photos and story to a major Northwest newspaper. I suggested that we contact an official of the firm with whom I was acquainted to get those logs out of the creek as soon as possible. I was certain nobody in the company knew they had escaped from captivity, and this move would exhaust the available administrative remedies, as a lawyer would put it.

Brock Evans, the Club's Pacific Northwest lobbyist who went on to become their Washington man and then Audubon Society lobbyist, overruled my suggestion: "Why should we give that company a chance? Demographers show that the environmental movement may have only a few years of high public popularity left. We have to win all the fights we can while the winning is good." So within a few days we got legislation by headline. Obviously no one in the Sierra Club gave a damn about the creek or the resource out of place. All that mattered to these "saviors of the environment" was their own political clout. Extend this across the top 10 pressure groups and you have a good case of moral bankruptcy making the world safe for hypocrisy, as the late scientist Robert White-Stevens put it. It is self-interested power, not simply environmental quality, that these leaders are after.

ETHICAL CONUNDRUMS

But this degraded ruthlessness does not square well with the seeming lovingness of the environmental ethic. How can this be? Listen to that ethic in its most reasonable and succinct form as the Shakertown Pledge from the American Friends Service Committee of 1975: "I commit myself to lead an ecologically sound life . . . we pledge that we will use the earth's natural resources sparingly and with gratitude. This includes the use of land, water, air, coal, timber, oil, minerals, and other important resources. We will try to keep our pollution of the environment to a minimum and will seek wherever possible to preserve the natural beauty of the earth."

It seems quite sensible and even uplifting. How can this good intention lead to bad deeds? Notice: nowhere is there a mention of their fellow men. There's the rub. They will use the timber and oil, but they won't recognize that a fellow human has to **produce** it for them. They will use the land, but never acknowledge that some fellow human has to manage it for nature preservation, or finance its commercial development, or forego opportunities if it is designated Wilderness. They're willing to breathe the air and drink the water, but unwilling to even think of the people who pay through the nose to clean it up from being sullied in the process of feeding, clothing, and sheltering us all. They will use all these things with gratitude, but none of that gratitude goes to the industrial benefactors who make it all available for sale to them. The environmental ethic, in short, is an elegant mockery of working humanity. When it comes to people, this ethic simply blanks out.

I think this is symptomatic of the whole environmental revolution. We have come to a point reminiscent of wildernist Aldo Leopold's complaint in the 1949 *A Sand County Almanac*: "Your true modern is separated from the land by many middlemen, and by innumerable physical gadgets. He has no vital relation to it." Since Leopold's day, affluence, wilderness ideology, and the automobile have brought million back (or forward) to the land. But there's a parallel problem now: Our "true moderns," mostly service sector employees, are separated from the realities of resource extraction and conversion by many comfortable middlemen and by innumerable ideological gadgets. They have no vital relation to the productive industrial source of their richly endowed lifestyles. Moderns are blind to the virtues and values of industry because they are immersed in them; it was never a fish that discovered water. So just as in Leopold's day, when the time was right for a "land ethic," so today the time is right for a "civilization ethic."

AN ETHICAL ETHIC

A fitting cornerstone for such an ethic was laid in the words of political scientist Ronald Inglehart's *The Silent Revolution*:

> The course of wisdom is difficult. It requires both a warm heart and cold reason. Perhaps the whales are sacred and the trees are sacred and the land is sacred. But mankind is sacred too and unique in that he seeks salvation and uses tools. If he abandons either, he abandons humanity.

Some may disagree with Inglehart's willingness to grant creatures the status of "sacred," feeling that only God is sacred—worthy of worship—but even these people agree that mankind seeks salvation and uses tools and that "if he abandons either, he abandons humanity." So how do you build an ethic wide enough to embrace both salvation and tools? In *Science and the Structure of Ethics*, Abraham Edel found that every ethical system is based upon some particular way of viewing existence that contains basic assumptions about the world and human nature; this he calls an Existential Perspective, or EP, by which he means "a view of existence."

The existential perspective of a "civilization ethic" is not too difficult to assemble. It must start by assuming that the world is real, that it has an existence outside our minds or wills, that it operates on natural laws which may or may not have been commanded by a Creator depending on your religious convictions, that mankind has a right to be in that world, that we are as natural as anything in the universe, that we have the right to save ourselves from any threat, whether from natural forces or our own folly. Such an ethic would recognize that man is a complex being that must deal with emotions and rationality, and that the urge to create is as basic as the urge to use language or to feel emotions. It would recognize that human creativity has always used tools—technology—and *that* inevitably results in modification of the environment.

Because the universe did not come with a set of instructions and because we can't reliably see the future, we make mistakes in modifying the environment. But because we possess emotional insight, rational thought and the intuitive creative urge, we have the ability to learn, to use better foresight, to correct and set right our mistakes, and ultimately to responsibly manage the earth forever. (Definition: **manage**—to treat with care. *Webster's Third New International Dictionary*.)

With this Existential Perspective in hand, we might state the ethic of civilization something like this:

I pledge to help produce, to wisely use, and to preserve the resources of my civilization, its food, fiber, minerals, energy, education, government, institutions, economic security, and ideals, and where possible to protect its wildlife and natural beauty. I will respect the earth's ability to support civilization, striving to minimize pollution and disruption of the natural world. I will respect the technical processes that are essential to the operation of civilization, knowing that they are as interdependent as any ecosystem. I will strive to recognize the benefits and limitations of civilization in relieving human misery, and in opening opportunities for a life of security, fulfillment and the refinements of aesthetic appreciation for its citizens.

THE LIVING IDEAL

And if you look carefully in that statement you will find the deepest values of American industry. This is as close as I can come to distilling the profound dignity of the worker's begrimed face; this is what I found in the forester when he walked across that land with me, his eyes on a horizon dec-

94

ades in the future, proud of his genetically enhanced, seed-orchard-produced, tree-nursery-grown, hand planted, pre-commercially thinned, fertilized, brush-controlled, and scientifically harvested forests; this is what gleamed beneath the mill operator's enthusiasm for waste-reducing thin-kerf saws, new board types, and other mileposts on the road to 100 percent raw material utilization; and this is what stirred the grizzled bullbuck one rainy Sunday in a remote bunkhouse to regale me with logging yarns full of ecological insights worthy of a great teacher and brimming with a sense of natural beauty reflected in pungent and striding language worthy of Walt Whitman.

The possession of a civilization ethic will not win our ecology wars. But it will do this: if you hold it up in public, in Congress or in the courtroom, it will make the motives of its attackers clear for all to see. It may even bring around some of those who embraced anti-technology and anti-civilization without realizing it, leaving the people-haters and civilization-killers standing alone.

Perhaps everyone in American industry doesn't live up to the civilization ethic, or any other, but those individuals and firms known who they are—and so do all the ethical industry folk. I have come a long road from narrow conservation supporter to the broader path of free enterprise advocate. I have seen industry take too many bum raps. To my friends in private enterprise, I hope this brief analysis has dealt a few good cards into your hands, even if you're not perfect. For all your warts and freckles, you're good people. Somebody else can come to bury you—I'll stick around to praise you.

6

How Environmentalist Propaganda Works

prop·a·gan·da. noun **Dissemination of ideas, information, or rumor for the purpose of helping or injuring an institution, a cause, or a person.**

NOWHERE DOES THIS DEFINITION SAY that propaganda is lies—rumors, true, but rumors are merely unverified reports, not necessarily lies. Yet American industry routinely makes the classic mistake of assuming that all environmentalist propaganda is lies and thus adopts the wrongheaded posture of defensive rebuttal. While much environmentalist propaganda is lies, the essence of its power is, as the definition points out, ideas, and ideas cannot be fought by mere rebuttal. As Victor Hugo said, nothing is so powerful as an idea whose time has come. If we are to realistically combat environmentalist propaganda, we must understand its ideas, why their time came in the course of American history, and we must learn to fight ideas with ideas.

THE INVISIBLE IDEA

One of the most powerful ideas in the environmental movement is not recognized as an idea at all: the non-profit volunteer membership struggle group. Without it, there would be no environmental movement. John Muir realized the political power of the struggle group when he founded the Sierra Club in 1892 and said of America's forests: "God has cared for these trees, saved them from drought, disease, avalanche, and a thousand straining, leveling tempests and

97

floods, but He cannot save them from fools—only Uncle Sam can do that." (Is it silly to note that Muir here attributes more power to the government than to God?)

Today's massive environmental bureaucracy proves Muir's prophetic gifts, for struggle groups have propagandized Uncle Sam into creating an "Ecoligarchy" with nearly dictatorial powers—the Environmental Protection Agency, for example, is one of the largest bureaucracies in the entire government.

Volunteer struggle groups have numerous built-in advantages as mechanisms for conducting propaganda campaigns. The first and most obvious is that the public can *join* volunteer membership groups, can be a recognized part of a movement, can work with the leaders of a cause. For many years industry had no similar hard-core citizen support groups that the public could join, and industry suffered accordingly.

Volunteer membership groups also can function adroitly in all three combat zones: public opinion shaping, legislative lobbying, and litigation. Their organizational structures allow plenty of freedom for quick decision-making at all levels. The national organization of any group such as the Sierra Club or Friends of the Earth is only the top of a large pyramid where general policy originates—beneath it are regional chapters, each made up of many local groups. Every level has clearly defined authority to release its own media material, to organize its own "telephone trees" to urge members to flood Congress with letters, to harass industry and government administrators with challenges and lawsuits, and to politically educate members and sympathizers.

But there are more advantages to membership struggle groups than structure. Non-profit groups command respect for their anti-establishment stands by invoking altruism, which even industry's trade associations cannot do for their own cause. Propagandists know that struggles which transcend personal concerns are likely to be more radical and merciless than conflicts over individuals' problems, and so seek some lofty ideal in their sloganeering. Communications sociologist Hugh Dalziel Duncan said in *Symbols In Society*, "As we plunge into battle, we cannot fight long and hard unless we fight in the name of some great principle of social order."

And Lewis Coser noted in *The Functions of Social Conflict* that "the consciousness of speaking for a superindividual "right" or system of values reinforces each party's intransigence, mobilizing energies that would not be available for mere personal interests and goals." The ability to rally around ideals is a main strength of struggle groups; it not only buys a good public image, but also generates more dedication, commitment, and energy among its adherents. This is one reason why employee groups have never been terribly effective in fighting for the industry side of issues: they're working in their own individual self-interest and they know it. In fact, so is the other side, but they don't know it: environmentalists are working to enforce their personal lifestyle preferences, their **mindstyles**, if you will—which is in their individual self-interest. But it is easy to fool yourself about such personal preferences by claiming that your mindstyle is some "great ideal" or "superindividual right." If you're fighting to preserve something concrete such as your job, it's harder to be such a fool.

STRUGGLE GROUPS

Building on these human psychological traits, environmentalist struggle groups can adopt a forceful and aggressive propaganda program (strategy) and run it freely with whatever projects prove workable (tactics). They are not presently constrained by truth-in-advertising laws or anti-trust laws, by liability for delays in developments, nor are they under any legal obligation to negotiate settlements as are labor organizations or moral obligation to abide by any professional code of ethics.

The most successful strategy of environmentalism was devised by the Wilderness Society and the Sierra Club in the years just before 1960. They recognized the futility of fighting for one wilderness area at a time and began the campaign to officially designate Wilderness as a "something." The result was a law that would provide a catch-all for ever-increasing acreages while putting the forest, grazing, petroleum and mineral exploration sectors on the defensive: the

Wilderness Act of 1964. Implementation of this brilliantly orchestrated strategy proved the effectiveness of the struggle group in bringing about political change—and conversely, the increasing weakness of the corporate business firm in defending itself in such battles.

Since 1964, more than 3,000 American environmental groups have sprung up, according to the *World Directory of Environmental Organizations* and the *Environmental Protection Directory*, totaling over 5 million members, or about 2 percent of our entire population. But this tiny 2 percent has lobbied dozens of laws through Congress, demanded and got scores of bureaucracies, agencies, councils, and committees created in the government, and prosecuted more than 500 lawsuits worthy of mention in law school texts. The changes in American government created by this tiny but vocal minority are truly astounding, a tribute to the power of the struggle group's structure and sociology as much as to its message.

Political awareness is the hallmark of the environmental struggle group. The lofty nature ideologies of Thoreau, Muir, and Aldo Leopold, among others, contain a tough-minded and aggressive determination to force social and political change. *The Sierra Club Political Handbook*, edited by Eugene Coan, is a masterpiece of practical politics, giving Club members pointers in seven vital areas. The handbook first explains how Congress works, emphasizing how committees can make or break legislative proposals. It next tells how to design and draft legislation correctly for the best reception by Congress. It then spends two chapters telling how to influence Congress, one on lobbying, which goes into incredible detail on backgrounding the club's Washington staff, the pitfalls of tax deductibility, how to avoid illegal actions, and how to successfully court Congressional aides; and another on the grassroots, showing step by step how to organize a devastating local political campaign, how to establish contact with important politicians, how to keep their attention, get their commitment, make them accountable for their positions, and how to dominate hearings with carefully rehearsed witnesses.

The handbook outlines practical letter-writing tactics,

warning "Don't mention that you are a member of the Club or that you are part of a letter-writing campaign. A letter from a "concerned citizen" is much more effective." In blunt terms, the handbook instructs, "The election opponent of a congressman may be delighted to take a stand if the congressman has not, or took the wrong position. To have his anti-environmental stand challenged in a campaign frequently forces a congressman to change his mind."

In its fifth chapter, neglecting no possibility, the book tells that, "It is a mistake to place too much emphasis on the legislative process," pointing out the effectiveness of hassling administrative agencies with challenges and lawsuits. Chapter six notes the importance of political education to the club's success, while the final chapter details how to manipulate the media, how to get on the wire services, and how to get free time on radio and television.

Making environmental propaganda work also requires legal astuteness of a high order. Where corporate attorneys are oriented toward minimizing the liability and exposure of their firms and lean toward settlement and timidity, environmentalist lawyers are furiously ambitious, not to follow the law, but to create it, to carve out new fields of theory, to innovate using the tools of lobbying and litigation, to create a whole new corpus of statutory and case law reflecting environmental ideologies. And they've done it. It is primarily the environmentalist struggle group and its attorneys that have made environmentalism an idea whose time has come in modern America.

REGULATIONISM

It is important to know exactly how they did it and what historical forces they built upon. And here we come to the story of regulationism, which in my analysis goes together with wildernism (examined in Chapter Two) to make up environmentalism as a whole. The drama of regulationism will take us from America's original zealous protection of individual rights and free enterprise to the regulation state of today with its dogmatic and coercive utopianism.

Regulationism rests on three cornerstones: public ownership of resource lands, agencies with the power of law to direct activities on public and private lands, and populist suspicion of corporate power and scope. From the earliest days of the United States national policy was to dispose of all public domain lands and put them into private hands, yet the Constitution is silent on the question. Likewise, regulatory agencies are not sanctioned by any provisions of the Constitution; original American policy was that government might act benevolently but never restrictively, never interfere with or control private business affairs. And populist suspicion of the corporation, which may seem quite recent, has in fact been with us since Jefferson's time, and is the seed from which all American regulationism grew. James Sullivan, attorney general of Massachusetts, said in 1802, "The creation of a great variety of corporate interests . . . must have a direct tendency to weaken the power of government."

Of the Constitution's silence on the question of disposing of government lands into private hands, some modern-day conservatives interpret Article I, Section 8, Subsection 16, as limiting the government's power to own lands to a ten-mile-square district—now Washington, D.C.—and to "all places purchased, by the consent of the legislature of the State in which the same shall be, for the erection of forts, magazines, arsenals, dockyards, and other needful buildings. . ."

Unfortunately, such well-meaning advocates of liberty ignore several harsh realities, among them Article IV, Section 3, Subsection 2 of the Constitution: "The congress shall have power to dispose of and make all needful rules and regulations respecting the territory or other property belonging to the United States." In practical terms, this has meant that congress can buy or sell any kind of property in any quantity, including land. Nothing in the U.S. Constitution suggests that forts, magazines, arsenals, and so forth, are the *only* specific properties the United States may own.

More unfortunate, these advocates of liberty also misread the whole nature of our Constitution or any constitution. A government's power to own or to take land flows from the power of eminent domain, which is a pre-constitutional power, not something granted by a constitution. As Schnidman, Abrams and Delaney said in *Handling the Land Use Case*, "Eminent domain or condemnation is the power of the sovereign to take private property for public use without the owner's consent. The power of eminent domain is inherent in every sovereign government and lies dormant until the legislature, by specific enactment, specifies how the power is to be used. The power is not given to government by the Constitution; rather, the Constitution is a limitation on the use of the power." What a weak instrument our Constitution is to protect private property will become evident.

The Constitution comments on general government ownership of land in the Fifth Amendment, which states, ". . . nor shall private property be taken for public use, without just compensation . . ." and in the Fourteenth Amendment which states, ". . .nor shall any State deprive any person of life, liberty or property, without due process of law. . . ." Put in different terms, the Fifth and Fourteenth Amendments themselves merely state that federal and state governments can take anything they want from anybody they want as long as they first pass a law or a court judgement taking someone's property and then pay for it. Thus, the actual security of private property in America is on much shakier legal ground than most of its citizens realize. This means that our vigilance against government abuse must be much greater than most of our citizens are willing to give.

CORPORATIONS: DARTH VADER IN A THREE-PIECE

Let us deal with the three cornerstones of regulationism in reverse order, taking populist suspicion of corporations first. The corporation became a popular form of business organization in the early 19th century; in all of the 18th century, only 335 corporate charters were issued in America. But the new nation that had carefully built checks and balances into its Constitution also feared unbridled power in large landowners and dynastic wealth, which the corporation seemed to imply (most corporations at that time were transportation monopolies, banks, or insurance companies— aggregations of "capital" that represents the "few" against the "many"). Everyone knew that the "artificial person" of the corporation was really run by normal human beings, but the word "soulless" appears constantly in anti-corporation propaganda of the time because corporations do not die and they have no ultimate size, very unlike a natural human being. Many free-enterprisers also fail to realize the fact that corporations are entirely the creature of the State: a government is the only power that can create a corporation. The private sector cannot give a company limited liability or tax advantages or an indefinite lifespan. A completely *laissez faire* economy could contain no corporations, since *laissez faire* rejects any government role in the economy. Yet the corporation has been the most efficient instrument for creating vast wealth ever devised.

During the early nineteenth century, people worried that the wit, skill, and malevolence of many men, unrestrained by considerations of family or morality, would come to political power by the sheer economic success of their corporations. There was little interest among anti-corporationists in forming competitive corporations to check and balance this feared power, just as today the anti-corporation environmentalists are not interested in forming corporations of their own that will mine and drill and harvest timber properly and manufacture without pollution. They were simply *against*. They didn't want to do anything right, they simply didn't want others to do anything wrong. And that set the stage for the later growth of regulationism as a governmental power.

EXPLODING TECHNOLOGY

Even so, the government protected free enterprise with great fervor, believing that the enlightened self-interest of the entrepreneur sufficed to guarantee the public's interests. This happy attitude was changed by a technological innovation: the steam engine. Steam power was transforming American culture in the early 1800s and making westward expansion a practical national policy, which delighted most citizens, even though the belching locomotive infuriated Thoreau as it intruded upon his nest at Walden. But steam engines, particularly on passenger and cargo boats, had the disquieting habit of exploding with horrible loss of life and property.

104

Jefferson had advocated in his 1801 inaugural address "a wise and frugal government, which shall restrain men from injuring one another, shall leave them otherwise free to regulate their own pursuits of industry and improvement." Prior to the steady stream of boiler explosions on river boats, only the freedom part of this dictum received much notice, but soon the restraint part began to take precedence.

As explained by John G. Burke in *Bursting Boilers and the Federal Power*, a small group of technically knowledgeable people formed the Franklin Institute in 1824 to investigate steam boiler explosions and became prime propagandists for federal intervention and regulation. In that same year, in the U.S. Supreme Court case of *Gibbons v. Ogden*, concerning the navigation rights of steamboats (unrelated to boiler explosions), Chief Justice Marshall threw open the door to regulation by ruling that the power of Congress to "regulate commerce with foreign nations and among the several states" (Article 1, Section 8, U.S. Constitution) is virtually absolute. But even with this landmark decision and the efforts of pro-regulation propagandists, it was more than a decade before Congress took the awesome step of regulating private enterprise, and when it did, the Franklin Institute had won. The grandaddy of all American regulationism was the Steam Boiler Act of 1838, containing three pages of rules designed to protect the safety and health of the general public from boiler explosions. An August 30, 1852, amendment of this law created America's first formal regulatory agency, a nine-man Board of Supervisors appointed by the President to enforce steam boiler regulations by investigating infractions and accidents with the power of subpoena. This drift away from basic American policy was seen as being so harsh that Senator Robert F. Stockton of New Jersey asked during bebate, "What will be left of human liberty if we progress on this course much further?" Good question, but it is important for modern citizens to know that regulationism began with a health and safety issue.

O GIVE ME A HOME

While the birth of regulationism was taking place in faulty steam technology, its future growth was being prepared in public land policy. The disposal of public domain lands was initially intended simply to raise revenue and to reward Revolutionary War veterans, but soon became a tool to encourage westward expansion and settlement through the pre-emption laws of 1796, 1820, and 1841. Congress gave enormous land grants to each new state, particularly under the Morrill Act of 1862, which gave states land to establish "colleges for the benefit of agricultural and mechanic arts." Land grants were given by the Homestead Act of 1862, railroad grants, and the Mining Act of 1872. But 1872 also saw the first true "conservation" law, establishing Yellowstone National Park. As the conservation movement began to gather force, it was realized that keeping lands in federal ownership was the key to control, just as Muir later stated.

Nonetheless, disposal continued apace: the Timber Culture Act of 1873 gave 160 acres to anyone who would plant trees on 40 of them, and the Desert Land Act of 1877 sold cheap land to whoever would irrigate it within 3 years. Yet it was not just conservation sentiment that finally reversed the policy of public land disposal; scandals were just as powerful. Frauds in obtaining the lands were legion, and once the lands were in private hands, poor husbandry gave the conservationists something to howl about. In the final decade of the 19th century, conservation forces gained strength, yet their first major victory, establishment of the forest reserves, was not won by fair debate in Congress. It was gained by the stratagem of a virtually unnoticed rider to an act repealing the timber-culture laws, a rider improperly added in a House/Senate conference committee and not referred back to the originating committees for their consideration. It is amusing to note that this back-door authorization of what eventually became the Forest Service is today known by the euphemism, "The Creative Act of 1891," or "The Forest Reserve Act," when nearly every Congressman at the time thought he was merely repealing an old 1873 law.

As the conservation movement arose during the last quarter of the 1900s, so did antibusiness regulatory sentiment. The Windom Committee of 1874 investigated the railroad industry and invoked the old Steam Boiler Act of 1852 to justify further regulatory legislation, resulting in the Interstate Commerce Act of 1877 and the Interstate Commerce Commission (1887). By 1890 only a small minority thought that government should not regulate business.

Scandals helped pass the Sherman Anti-Trust Act of 1890, and conservationists did not miss the significance of federal power in regulating natural resource projects. The Organic Administration Act of 1897 favored "withdrawal" of public lands from private sale, establishing national forests "for the purpose of securing favorable conditions of water flows, and to furnish a continuous supply of timber for the use and necessities" of Americans, thus clearly expressing the new awareness that as the frontier closed, land and forests would become increasingly valuable. The Transfer Act of 1905 duly established the Forest Service in the Department of Agriculture to regulate forest use, although originally only economic uses were chartered by Congress—the multiple use idea was many years away. The Pickett Act of 1910 further weakened the public land disposal system by authorizing the President to make "temporary" withdrawals for a variety of purposes. Regulationism got a further boost in 1914 when the Clayton Act tightened "restraint of trade" provisions of anti-trust and the Federal Trade Commission was created to regulate a wide range of business practices. The creation of the National Park Service in 1916 put preservationism on the federal map, and in 1924, the world's first official wilderness area was founded in Gila National Forest in New Mexico. Regulationism was then a workable tool of the conservation movement.

The Taylor Grazing Act of 1934 spread regulationism to grazing lands to make it more difficult for public land to pass into private hands. But it was not until 1964 and the act that created the Public Land Law Review Commission (PLLRC) that Congress made it clear that the long-standing statutory preference for disposal was at an end, and that "maximum benefit for the general public" was the new goal. The PLLRC in turn recommended that "environmental quality should be recognized as an important objective of public land management" on the 725 million acres of remaining public land, one third of the nation's total area. That objective was made into law by the National Environmental Policy Act of 1969, which tied together federal control of the public lands, the regulation of business, and populist suspicion of corporate power and scope. Thus America became a thoroughgoing regulation state.

So now we have the new breed of "regulation man," as he is called by Dr. Madsen Pirie of the Center for Constructive Alternatives, devoted to enforcing protective measures "for life, health and property" (Dr. Pirie ponders "wasn't **liberty** on that list once?"). Regulation men such as Ralph Nader believe that anyone who opposes their regulatory methods and philosophy opposes their avowedly pure intentions, the essence of the totalitarian mentality. We are faced with a brigade of "safety and health fascists," as the late master strategist Herman Kahn called them. The power of these advocates poses a clear and present danger to the freedom of our open society that must not be taken lightly.

LEGITIMACY

It would be a mistake to believe that environmentalists came to power only by legal and political astuteness, or organizational expertise, or by manipulating zealous memberships, even though these are prime factors. In the final analysis, all legitimate power rests on the acceptance of a lifestyle. But by clever and persistent propaganda, and the liberalizing effects of affluence and upward mobility, environmental advocates can gradually change public perceptions of the world, as this brief history shows, and step by step gain acceptance for the lifestyles they advocate. American industry should not respond to these changes with blind anger, but with curiosity and study, because the environmentalist program *has worked*. Only with a clear grasp of its methods and principles can it be defeated.

One environmentalist propaganda method has baffled workers and managers and deserves comment: the insistence on economic efficiency for management practices on federal lands. Many in natural resource industries have been stunned by this cost/benefit incursion by environmentalists on industrial economic turf. It they had kept up on their reading, they could have seen it coming: Marion Clawson said in his 1976 *The Economics of National Forest Management* "an approach based upon the economics developed in this paper will do more to help [preservationists] attain what they want from national forests, than will exclusively emotional appeals." Environmentalists did read it, and took the advice.

AMENITY ECOMONICS

The most concise statement of environmentalist economic wisdom is contained in *A New Reality: Timber Land Suitability in Oregon National Forests* by Randall O'Toole of Cascade Holistic Economic Consultants, or CHEC. Although the study is regional, its tactic is applicable throughout the United States, as it relies on federal laws, harking back to Muir's Uncle Sam dictum. CHEC is a non-profit consulting firm providing services at or below cost to environmental groups. *A New Reality* was sponsored by Oregon Wilderness Coalition, a band of struggle groups including the Sierra Club.

In a nutshell, the "new reality" is this: "that standing timber should not be harvested unless the land on which it stands is capable of efficient timber production." In other words, forest lands should be classified submarginal and put off limits to harvest unless the costs to regenerate and manage the land do not exceed the dollar return from the regenerated forest when it is harvested decades in the future, including interest. This is a test to see whether the reforestation investment would be better spent elsewhere. The economic method O'Toole uses is not so new: it's a fairly standard way to deal with regeneration decisions on industrial lands. What is new is O'Toole's conclusion that federal lands should suddenly use free market Adam Smith-type economics instead of continuing with Keynesian welfare economics—managing land for "the greatest public benefit"—and without eliminating the vast and costly management bureaucracy that the free market would never tolerate. It is merely a ploy to get "marginal" lands declared "submarginal" and prevent timber harvest on them. O'Toole's theory and calculations are lifted bodily from Fisher and Krutilla's seminal work *The Economics of Natural Environments*, which asserts that amenity values must be given equal and superior status with dollar values.

The key to understanding the new tactic is this: the calculations paste free market *methods* on what is not a free market—federal lands—with no corresponding disempowerment of the federal government. Private property interests in federal lands are never mentioned—timber sale contracts, proprietary water rights of cattlemen in springs and wells on federal gazing lands, and so forth. Private buyers and sellers are not given equal power footing with federal bureaucrats—it is simply not a free market situation. Nor is the immense overhead of the federal bureaucracy accounted for in costing out regeneration programs. In short, both the calculations and conclusion of O'Toole's approach are not economics, but politics.

Using O'Toole's calculations you could also come up with at least these two contrary conclusions: 1) as advocated by University of Washington free-market forest economist Barney Dowdle, the timber should be harvested for strictly economic reasons and the land allowed to regenerate naturally with no financial investment, or 2) the timber should be harvested to promote rural community stability and provide federal aid to small business (local sawmills). The *United States Government Manual* clearly states that among Forest Service objectives are "generation of forestry opportunities to accelerate rural community growth; encouragement of the growth and development of forestry-based enterprises that readily respond to consumers' changing needs." So by using O'Toole's numbers, you can arrive at valid rebuttals in both free-market and Keynesian welfare economic theory. The whole point is that economics alone proves nothing. Politics, on the other hands, proves anything the propagandist wants it to. This distinction must be fully grasped by American industry if it is to survive this new combat tactic.

At bottom, the Multiple Use-Sustained Yield Act of 1960 is the culprit. It was the first American law to deny economics. It says that "the needs of the American people" come first, and that the national forests shall be managed "with consideration being given to the relative values of the various resources, and not necessarily the combination of uses that will give the greatest dollar return." This clause undermined the economic use of federal lands and directly led to the present disaster of commodity production on federal lands being relegated to a residual use that is tolerated only after all other demands are met. That law must be repealed, along with every other "environmental" law, and new economic/ecologic laws enacted to reshape the federal domain into some semblance of rationality.

This is no idle dream. It can be done. It is possible to protect nature without destroying mankind's economy. It is possible to manage industrial civilization without destroying nature's ecology. One day we will wake up and see clearly that an economy is human ecology. We will learn that *it all costs*, the air, the water, the land, the commodities, the means of production, the social structure, the culture. Wilderness and nature preserves will no longer be treated as an unpriced good, nor will industry be crushed with regulations that yield no practical benefit. Reasonable economic/ecologic thinking will take the place of special pleading. The federal estate will in fact behave as part of the market mechanism. But as things stand now, the federal estate is merely a political playground for "post-gratification forgetting and devaluation." One day those sleepers will be forced awake. But only after we have stripped environmentalist propaganda of its mystique and clarified precisely its real-world consequences.

All this should reveal a great deal about how environmentalist propaganda works. The principle behind environmentalism may seem familiar, and it should. It is the same rationale that has attended the rise of empires throughout history: the end justifies the means. The claims are still the same: a better world. The goals are also still the same: totalitarian power. And it is still just as wrong.

7

Defeating Environmentalism

THE GOAL IN OUR ECOLOGY WARS should be to defeat environmentalism. That's an explosive statement, and it's important to understand what I mean by it—and what I don't mean. Environmentalism is so entrenched that the mere suggestion of defeating it quickens our pulse, widens our eyes and tenses our muscles. I'm NOT suggesting that we defeat the *environment*. Defeating environmentalism will not defeat the environment. Environmentalism is not the environment. Environmentalism is an institutionalized movement of certain people with a certain ideology about man and nature, specific people with a specific lifestyle/-mindstyle. Environmentalism does not necessarily have the environment's best interests at heart. Environmentalism is an institution with its own survival at heart. Environmentalism needs the environment. The environment does not need environmentalism.

A healthy outlook on the environment does not require that we adopt the views of environmentalism. We can love the earth and its community of life without hating technology, without wanting to destroy industrial civilization, without wallowing in an orgy of self-loathing and without drowning individual rights in a totalitarian "Ecotopia." It is the ideology, the destructive "**-ism**" in environmentalism that we should defeat. It is the excess baggage of anti-technology, of anti-civilization, of anti-humanity, of institutionalized lust for political power that we must reject, that society must overcome. No one will miss those things. The environment cannot be improved by those things.

113

Like all new ideas, this one must be pounded home to make sure we get it right. Let me repeat, wanting to defeat environmentalism does not mean that we are interested in defeating the environment. Quite the contrary. Renewable resource owners—farmers, ranchers, timberland owners— have historically protected their resources much better than any government, even considering the cut-out-and-get-out episodes of 19th century America and the Dust Bowl catastrophe of the 20th century. Most resource owners do not share the views of institutionalized environmentalism.

Environmentalists **always** turn to government to do their dirty work. But it is government that creates the "commons" of which Garrett Hardin wrote in his *Tragedy of the Commons*. Government is the problem, not the solution. Hardin himself recognized that "the tragedy of the commons as a food basket is averted by private property, or something formally like it." He added the "or something formally like it" because he is basically a statist and wanted to legitimize government ownership as being formally like private property, which it is not. Private property does not breed bureaucracies out for job security nor does it attract non-title-holding special interests out to allocate land use without paying for it.

Alternative Viewpoints

We have learned many vital lessons from the positive aspects of the environmental movement, but we should also learn to look beyond its negativism. Defeating environmentalism, as the thoughtful reader will already have realized, means taking on a more benevolent, more mature and more psychologically healthy vision of environmental advocacy, one not based upon increased government coercion. What would such a vision encompass? It would be broader, more holistic, and include mankind in its list of worthy organisms. It would care for human well-being and pleasure, both spiritual and material. It would affirm the individual rights of tool-using, salvation seeking humankind while nurturing a high regard for other life forms and the physical universe. It would not be so frightened of the "delicacy" of Earth's ecosystem.

A good example of such a worldview was provided by British scientist James Lovelock in his inspiring and challenging book *Gaia: A new look at life on Earth.* His basic thesis is that earth's biosphere is a self-correcting organism capable of dealing with most of man's intrusions—as long as we avert nuclear war. Despite its literary invocation of the Greek Earth-goddess Gaia, Lovelock's intriguing postulate came from deep study of Earth's various systems, land, sea, air, living creatures. He asserts that the facts do not support the conclusions of many ecologists about steady-state nature and steady-state economies. Concerning the air, Lovelock said:

> The chemical composition of the atmosphere bears no relation to the expectations of steady-state chemical equilibrium. The presence of methane, nitrous oxide, and even nitrogen in our present atmosphere represents violation of the rules of chemistry to be measured in tens of orders of magnitude. Disequilibria on this scale suggest that the atmosphere is not merely a biological product, but more probably a biological construction: not living, but like a cat's fur, a bird's feathers, or the paper of a wasp's nest, an extension of a living system designed to maintain a chosen environment. Thus the atmospheric concentration of gases such as oxygen and ammonia is found to be kept at an optimum value from which even small departures could have disastrous consequences for life.
>
> The climate and the chemical properties of the Earth now and throughout its history seem always to have been optimal for life. For this to have happened by chance is as unlikely as to survive unscathed a drive blindfold through rush-hour traffic.

Lovelock's clearsighted vision of a self-protecting Earth managed for ages by self-knowing human stewards slaps the doctrinaire environmentalist vanguard squarely in the political philosophy:

Pollution is not, as we are so often told, a product of moral turpitude. It is an inevitable consequence of life at work. In a sensible world, industrial waste would not be banned but put to good use. The negative, unconstructive response of prohibition by law seems as idiotic as legislating against the emission of dung from cows.

BLENDING ECOLOGY AND ECONOMY

So the underlying motivation for defeating environmentalism emerges powerfully and positively as a whole new philosophy of life: to recognize that the ecology of humanity and the economy of nature are entwined and survival-prone. In a more technical sense, ecology and economy are related but not identical; their relationship is exact but largely unexplored. The bridging concept is "econosystem." Ecosystems and econosystems are two aspects of the same total biosystem, both demanding respect and protection. They are alike, different and related, as alike, different and related as two sides of the same coin.

Econosystem is a new and crucial idea. Like the notion of an "ecosystem," "econosystem" is a conceptual model to help explain existence. In brief, an **econosystem** is human ecology. "The" econosystem is the entirety of interrelationships between human beings and the physical universe, including all other organisms. A basic premise of econosystemic thought is that humans and our disequilibria are as much a part of the biological construction we call the biosphere as the birds and the whales. While we cause havoc in some places we serve as a potent survival force in others. Just as viruses differ distinctively in their survival modes from bacteria and just as plants diverge characteristically from animals, human survival patterns contrast with those of all other living creatures in their own unique ways.

An econosystem is different from an ecosystem in that human beings, which create econosystems, **characteristically think rationally and make purposeful choices** about their interrelationships with existence. Members of non-human ecosystems do not do so to such an extent or so characteristically. Rational thought and purposeful selection define econosystems. Ecosystems are defined by random motion (physics) and natural selection (biology). The scientific implications of this definition are profound: the econosystem—human ecology—is not just another example of natural ecology, it is a qualitatively and quantitatively *different kind* of ecology, yet still embedded in the biological construction called the biosphere. Interestingly enough, an econosystem could be depicted on a standard input-output chart such as those used by economists and ecologists. See Walter Isard's highly technical study, *Ecologic-Economic Analysis for Regional Development*, for an early effort in that direction.

The differences between the laws of ecosystems and econosystems bear some resemblance to the differences between the laws of physics and the laws of biology. In physics, entropy is an absolute given: energy flows only from higher states to lower states, from higher degrees of organization to lower degrees of organization. Yet in biology we find increasingly complex degrees of organization—living organisms—which appear to be local violations of the general law of entropy. However, even within this diversity there appears a unity at the deepest level: although the biological world gathers and binds energy in higher degrees of organization, no energy in living creatures flows from lower states to higher states. The conservation of energy is maintained even where life organizes at its most complex.

The biologist had to seek and discover new and specialized scientific laws to explain biological phenomena, laws such as natural selection, the genetic role of DNA, and the biogeochemical modes by which living matter binds energy into complex structures. Evolution by natural selection does not concern the physicist: the natural history of kangaroos follows much different rules than the natural history of star-systems. Similarly, the social scientist could not work if constrained only to the scientific laws found in physics and biology: societies and minds cannot be understood by applying to them the laws that govern kangaroos or starsystems. Each discipline must work out the scientific laws appropriate to its own subject matter, even while realizing that at root there is probably a unifying bond that cements all knowledge together (don't ask for it yet: philosophers of science—creationists, evolutionists, realists, idealists, positivists, dialectical materialists, *ad nauseam*—are still shooting it out over that long-sought unified theory of all knowledge).

Thus it should come as no surprise to find that, even though ecosystems and econosystems are two different aspects of the same thing, the human econosystem operates on distinctive principles not found elsewhere in the biosphere. For example, even though the human econosystem appears to be a closed system totally embedded in the earth's biosphere, in fact it displays a remarkable ability to break out of nature's matrix. Rational thought and purposeful choice give humans the power to open the secrets of natural systems to a substantial degree. Humans can alter natural ecosystems to favor human survival, to release new forms of energy, to dominate all natural ecosystems, to learn the techniques of natural ecology, and ultimately to escape the limits of "nature" (but not the limits of "natural law"). "Natural" ecosystems have experienced a degree of this dynamism in the evolutionary thrust toward more and more complex species, over time breaking through old limits of structure, behavior, habitat and specialization. Humanity appears to specialize in breakthrough.

MASTERS IN SPITE OF OURSELVES

Human econosystems commonly push natural ecosystems in particular places until a breaking point is reached where a particular econosystemic method fails—frequently with temporarily disastrous results—but then rational thought and purposeful choice return to the failed problem with new methods to push natural ecosystems beyond the old breaking point until new limits are reached. The resulting never-ending outward spiral of power and knowledge appears unique to human beings.

The environmentalist vanguard takes a peevish and schizophrenic view of these disturbing truths: on the one hand, they deny that humanity has such breakthrough powers, and other the other hand they object to the everyday human use of such powers. But indeed these breakthrough powers exist. The results are all around us. Examples are increased human populations made possible by the revolutionary creations of agriculture, civilization and industrial production. Human beings are among the most successful organisms on earth by every biological measure: geographic range,

growth, size and health of population, resistance to catastrophe, adaptability, survival potential. Environmentalists
perceive these facts as an unmitigated evil—an attitude
which in itself is an unmitigated evil.

The notion of "econosystem" is a new paradigm. With
further development, it could even point to a new total
worldview, and perhaps advance new explanations of how
everything works. It has some interesting implications about
teleology. It can be looked upon as a more precise version of
Pierre Teilhard de Chardin's "noosphere" concept, but without the mystical overtones, or as a more inclusive version of
Ludwig von Mises' "praxeology" concept, but with a sys-
tems-theory understanding of society: the econosystem is a
new kind of "natural force" in the sense of being a function of
billions of human decisions interrrelating every day—
between man and nature—*as a system.* Econosystemic
thought could be called the Compatible Disequilibrium
Paradigm, indicating that the tensions between "man" and
"nature" are themselves a source of tremendous new biological survival power for both man and nature—a kind of
ultimate *denouement* that encompasses smaller localized
incompatibilities, such as man's tendency to push marginal species to premature but inevitable extinction, within
the larger inclusive compatibility of Gaia's pro-survival
dynamism.

In practical application, econosystemic thinking would find
techniques for weighing the inherent survival potential of an
endangered species before giving it special protection,
whereas an ideological environmentalist would make others
pay unlimited amounts of time, energy and money to protect
all species for quasi-religious reasons (all life is "sacred").

The econosystem concept demands not only a technical
development far more detailed than outlined here, but also a
rethinking and reworking of all past philosophies, tasks for
other books and other workers. But the seeds are there. They
bear the genetic code of a revolutionary vision of
untrammeled nature and untrammeled man, of a free economy and a free society that understands itself and nature,
and does not shrink from *managing* itself and nature. It is a
philosophic outlook of acceptance: it demands no excuses for

humanity. It demands no guilt for being what we are. It is able to appreciate humanity. It suggests that our outward reach, our innate starburst vigor, our urge to infinite expansiveness is our essence, and that our essence is *survival*. If you think the econosystem concept all the way through, you may find it suggests a vision of unparalleled beauty, culture, and social achievement. You may find it a vision of such magnitude that, properly developed, would repudiate all the enemies of the open society from Plato to Marx and imply vast changes in our lifestyles/mindstyles. It would imply, among other things, the elimination of environmentalist ideology because there would be no more need for it, and that is the absolute defeat of environmentalism.

THE TOOLKIT

But in the here and now, the question arises, "Can we defeat environmentalism?" We know that the environmental movement has arrogated immense power to itself in our federal "Ecoligarchy." We know that dozens of activists who have lobbied on Capitol Hill and sued in federal court have held and still hold jobs at all levels in the Environmental Protection Agency, the Council on Environmental Quality, the Departments of Interior, Agriculture, Justice, Commerce, plus the Occupational Health and Safety Administration and the Federal Trade Commission.

We have seen this breed of bureaucrat succeed in blocking virtually every avenue of future economic growth in America through chronic faultfinding. We know that in the U.S. Congress, the House has numerous pro-environmentalist members. We know that the case law of the past two decades in our federal courts of appeal and the U.S. Supreme Court overwhelmingly favors environmentalists. We also know, as Irving Kristol tells us in *Two Cheers for Capitalism*, that the "new class" of well-educated and affluent service-sector employees responds to a leadership out to determine America's social goals and to "dominate and define the society." This, says Washington University's Murray L. Weidenbaum, will result in a transfer of power from the managerial class to the bureaucracy—from the private sector to the government.

Some among us in the professions, such as Dr. James R. Dunn, president of the American Institute of Professional Geologists, recognize that the real motive of these "new class" environmentalists is not solely to protect the environment, but to forward their own political philosophies—no growth, anti-business, big-government, ultra-liberalism, socialism, communism, or what have you; environmentalists' declared political beliefs cut across a wide spectrum and most distressing, many environmentalists do not realize the political implications of their naturalistic philosophies.

Still others, such as Dr. H. Peter Metzger of the Public Service Co. of Colorado, also realize that once they have the political power, these environmentalists will bring about a new order: "For the first time in history, those in power will have decided that the goose—the wealth-generating machine, the economy—has laid enough golden eggs, and she's going to be retired . . . The economy will be shut down." They haven't managed that—yet. But the little tree-hugging David of 1960 has grown into the bristling political Goliath of the 1980s and '90s, replete with Political Action Committees and environmentalist lobbyists such as Brock Evans running for Congress in 1984 (he lost).

Against this background it is only reasonable to ask, "Can we defeat environmentalism?" My emphatic reply is, "Yes, we can." My only proviso is, "If we act." How then? American industry has warded off some legislative and litigative blows, but has suffered defeat after defeat at the hands of shrewd and sophisticated environmental activists and attorneys. Frustration and pessimism fill our daily lives. Obviously, what we have been doing doesn't work very well, so what will?

NEW QUESTION, NEW ANSWER

My answer may seem alien to some, but bear with the strangeness: we must combine our traditional approaches with the same activist techniques that have been so devastating in environmentalist hands.

I know that the word "activist" conjures up all sorts of negative images of shaggy-haired mobs of ultra-liberals and radicals that we want nothing to do with, but that's not what "activist" means at root. Activist simply means someone who is active in a cause, someone who gets things done. It's true that being activists falls outside of the experience of most hard-working Americans, but there is a good reason why I recommend it: you can only fight an activist movement with an activist movement.

Saul Alinski's *Rules for Radicals* advised activists to cause confusion, fear, and retreat in our ranks by applying the rule: Wherever possible, go outside of the experience of the enemy. The enemy was us, and they went outside of our experience during the 1960s and early '70s, all right. They appealed to emotions because we would never think of such stooping. They manipulated the media because we thought it was beneath our dignity to do so. They talked ethics and aesthetics because those subjects seemed to be beyond the scope of economics (they aren't, really, but that, too, is the subject for another book). They aggressively built a legal fence around our activities one stake at a time, a law passed here, a court decision there, because we couldn't even recognize what they were doing, it was so alien to us. They made a new world of discourse and set themselves up as its prophets. We were shut out because we didn't understand what they were doing—it was outside of our experience. Let me assure you as someone who has sat in activist councils of power, American industry can ultimately win only by expanding its experience and by striking back with its own brand of **activism**.

Expanding our experience means education. Whatever formal educations we may have had were certainly skimpy on social conflict and politics. But that's where hardball is being played in ecology wars. That's where the power is. So the first thing we need to do is educate ourselves to understand environmentalist sociology and environmentalist politics. We can't count much on our normal institutions to teach us. Either they don't know, or they're not saying. I've talked to business students all over America, and they're deeply concerned that their curricula are not preparing them for the real world they'll find upon graduation. We have plenty of forums in industry already—conferences, conventions, meetings—all we need to do is change their contents. Self education is a well-used environmentalist technique, as Earth Day, April 22, 1970, proved; its actual name, if you will recall, was The Environmental Teach-In.

We need speakers and writers who understand politics, and who will tell us what they know. Many have in the mid-1980s arrived and are well established: Charles S. Cushman of the National Inholders Association; David Dietz of Oregonians for Food and Shelter; Rob Rivett of Pacific Legal Foundation; free-enterprise-oriented non-profit organizations such as the Free Congress Research and Education Foundation, the Heritage Foundation, the American Enterprise Institute, the Pacific Institute for Public Policy Research and others. Here are a few of the things they're teaching us:

Drastic changes have come over American society since 1960. First, a sequence of doomsday warnings came, telling about the dangers of nuclear fallout, about harmful chemicals destroying wildlife and causing human disease, about unbridled population growth and the depletion of natural resources. These announcements dropped a pall of fear over our nation and electrified the public into intense resentment against industry for causing the problem. At the same time, the service sector was growing to vastly outnumber the combined agricultural and manufacturing sectors, and myriads of information-based industries grew up.

This work specialization brought about new social structures, loyalties and communication channels that eventually there was no longer any automatic sense of common interest within our nation, but rather a growing number of special interests each with its own life philosophy and modus operandi. This increasingly complex society was people with a younger average-aged citizenry from the World War II baby boom, and the affluence and security they grew up in led them to develop new personal values.

Today economic and security needs don't concern these younger people as much as they do the older generation, who were affected by the Great Depression and World War II. New needs for love, for a sense of belonging, for self-actualization, for personal autonomy, and high-level needs for intellectual and aesthetic gratification have become increasingly important to younger Americans. Environmentalist programs fill *all* these needs. If you will notice, American industry has not filled any of them, because only an activist program can do so.

But most importantly, as Roy Amara, president of the Institute for the Future, has noted: "There has been a great increase in citizen participation so that nearly all problems today become politicized." This increased participation has led to government by loggerhead—nothing quite gets done because the timid can veto the venturesome. And all of this social change has grown up in a new web of high intensity mass communication, with the media influencing events by setting the agenda of which issues will get public attention.

The key lesson to be learned from this instant sociology class was best said by Congressman Barber Conable: "This is an activist society." If our society has become activist, we'd better wake up and realize it. If our society has become activist, we'd better become activist, or we'll suffer the consequences. What are the consequences? If we don't take part in the political process, some person or group will actively influence the government in directions we won't like (haven't they?). It will do us no good to deplore this turn of events; it's what participatory democracy is all about. Law professor Jack Davies warned us in *Legislative Law and Process* that, "The legislative process is an adversary system in which silence is treated as acquiescence." If we don't speak up, we may lose the right to speak up. So the low profile may lead to the absent profile. Remember, the Greeks had a word for a private person who did not participate in politics, a word that has come down to us in English only slightly changed in meaning. The word? Idiot. In an activist society, we can't afford to be idiots.

CITIZEN ACTION

So how do we avoid it? What do we need to know in order to intelligently participate in politics? The basic facts are simple: 1) politics is a spectrum of activities, not just a single thing, and 2) ecology wars can only be won by taking appropriate action in all parts of that spectrum. Once we understand this spectrum, we need to act, to make sure we have mechanisms to operate in all its parts, particularly in the areas of developing public support and citizen activism. Former Chief Executive Officer Reginald H. Jones of General Electric made the point forcefully: "Business must develop a constituency—a body of hard-core supporters who will defend it in times of trouble and speak up for it in the debate over public policy. Business is the only institution that does not have such a constituency at the present time."

Jones' insightful remark was made in 1979 and is, fortunately, no longer true. In fact, even when he made the remark he was unaware that such a constituency was burgeoning from an unexpected quarter: movements calling themselves variously Women In Agriculture, Women In Timber, and Women Involved in Farm Economics (WIFE), among others, the first stirrings of a genuine pro-industry citizen activist movement. But its birth pangs showed clearly that any such constituency first has to fight industry almost as hard as it fights environmentalists. Trade associations in particular jealously guarded their power turf from the new interlopers. Leaders of industry firms also dismissed these organizations as "the little women" or "the ladies auxiliary." But when it became evident that their agenda was business only, to orchestrate public hearings, to sue in the courts, to lobby in Congress, to pressure administrators and in general to out-Sierra Club the Sierra Club, many in industry came to their support. Yet one benighted soul was heard to say, "We don't want to dilute our power in a proliferation of splinter groups!" But this neanderthal got his comeuppance from Betty Denison, at that time president of the Oregon Women for Timber, who retorted, "Gentlemen, what power? You haven't won a single completely favorable federal law from Congress since before the Multiple-Use / Sustained-Yield Act of 1960!" That may be a slight exaggeration, but it makes a real point: we need activists if we want a better track record.

Citizen activist groups allied with American industry are vital to our survival. By 1982 my wife Janet and I could find nearly 800 groups defending industries of various types, and we published their names and addresses in our *Directory of Pro-Industry Citizen Organizations: The Person-to-Person and Group-to-Group Pro-Free Enterprise Link.* We distributed 1,000 copies of that directory to help build the network.

We did it because we know that citizen groups can speak for us in the public interest where industry itself cannot. Citizen groups are not limited by liability, contract law or ethical codes. They can say bluntly what needs to be said without hiding behind corporate timidity or fear of offending customers. They can provide something for the masses to join, to be part of, to fight for, to satisfy that "need for a sense of belonging" that psychologist Maslow told us about. They can get on with aggressively building legal fences around environmentalists where industry and its associations can ill afford such a controversial role. They can network with other citizen activist groups such as Consumer Alert to form broad-based political coalitions of labor organizations, National Home Builders Assn., American Farm Bureau Federation, mining groups, cattlemen, manufacturers and on and on. They can keep up with local "street" politics and open doors that would be closed to industry among the citizenry. They can grab media attention by using every trick in the activist tote bag: creating dramatic confrontations, picketing, revealing ideological conflicts of interest with the public good.

In mid-1985 an angry group of some 400 citizens from Mapleton and Florence, Oregon, proved how effective activism could be. The Concerned Citizens for Western Lane County, upset by a National Wildlife Federation lawsuit that blocked all timber sales on the Mapleton Ranger District of the Siuslaw National Forest, hired buses to take them to Portland where the Federation was holding a huge conference. The outraged citizens picketed the hotel, invaded the meeting and backed Federation leader Jay Hair into a corner with pointed accusations of destroying our economy that he could not answer—all on network television.

This time the sign-waving protesters were defending industry. This time the environmental movement was on the defensive. This time the paid leadership of environmentalism—and Jay Hair gets $125,000 a year—ended up looking like jerks in the media. The Concerned Citizens shortly thereafter organized the shutdown of Florence, Oregon, a coastal resort and timber town, at the height of the tourist season for a four-hour demonstration of their plight. The media coverage was overwhelmingly favorable, with only one unreconstructed environmentalist TV news team panning the event.

Although Congressional leaders will never admit it, these demonstrations of pro-industry sentiment appear to have influenced efforts to release certain timber stands for harvest by the affected communities to replace the lost trees tied up by the National Wildlife Federation's lawsuit. This proves that activists can give big muscle to our expensive legal talent in lobbying crucial votes and unjamming reluctant Congressional committees by floods of letters and telegrams.

This is not an either/or proposition. Industry and citizen activist groups are mutually beneficial. Industry must come to support citizen activist groups, providing them funds, materials, transportation, and most of all, hard facts. Even the most wild-eyed environmentalist realizes that the days of the activist free-for-all are over. All the flashy tricks that still work in the media get you nowhere in Congress without hard facts.

As Dr. Jean Mater, author of *Citizens Involved: Handle With Care!*, says: "Industry can supply the facts on issues (backgrounders). Grass roots groups can supply the acts to focus public attention on the issues. Industry supplies the facts. Activists supply the acts." They're both essential. And this would cure the endemic disease of "talking to ourselves" that so many in industry bitterly complain about. Citizen activists will take our message where it belongs—not to conferences, but to the public—and they'll do it person-to-person, one-to-one. The power of one-to-one advocacy seems to be lost on industry, but it must never be underestimated.

As industry helps friendly citizen activist groups to grow, it must learn to think like an activist itself: it must become politicized. We must all learn how to become a grass root, even if the prospect doesn't thrill us much. Political education has to become a permanent part of industry effort. This detailed look at the political spectrum involved in defeating environmentalism can be the cornerstone of that effort.

ELECTING OFFICIALS

Many environmental battles are won or lost before they are even joined. They are decided when the public elects officials who have the authority to defend or wreck free enterprise for the next two, four, or six years. Political education programs that work to get sympathetic candidates elected have been legal for non-profit organizations under certain restrictions since 1975, and environmental groups take full advantage of that fact. Here are the basic tools of electing friendly officials:

The Usual. Contributions, doorbelling, arranging coffee hours, lobbying precinct committee-people, arranging speeches and rallies. Best done on a private basis to avoid branding the candidate as favoring some "special interest." You'll soon learn the political hypocrisy of environmentalist groups here: if they electioneer, it's the "public interest;" if you do it, it's a "special interest." The truth is, environmental organizations are self-serving special interests with a much more limited constituency than American industry.

Voting Charts. You can publish charts to show how any incumbent voted on key industry issues, with a percentage of correct votes displayed prominently. Just as the environmental movement singles out its "Dirty Dozen," pro-free enterprise groups can single out the "Dictatorial Dozen" who consistently vote for more and more government regulation and centralization of power.

Candidate Questionnaires. Designed to get reactions from incumbents and challengers to a set of questions on industry issues. Publicizing the finished questionnaire puts pressure on the candidate to keep campaign promises.

Accountability. Make the candidate aware he or she will be watched and publicized before and after the election. Make them into heroes when they do well and make sure they get bad press when they go the wrong way. Activist group newsletters should routinely publish the voting records of their local Congressmen.

Winning elections does not always give you the results you expected. Remember, conservative Richard Nixon created the Environmental Protection Agency out of thin air by a presidential reorganization order—Congress didn't create the EPA at all. However, it pays to screen campaign rhetoric for obvious warning signals. Jimmy Carter, for example, hoped "to challenge Ralph Nader in the future for the role of top consumer advocate in the country." Electing sympathetic officials can undo roadblocks in Congress such as Rep. John Seiberling (D-Ohio), a powerful figure in the House Interior and Insular Affairs Committee, who has arrogantly sneered, "if the Sierra Club doesn't like any given piece of legislation we're considering, it'll never get out of my subcommittee." And when we do manage to put a favorable candidate in Congress, don't think that your support will buy you specific performance. It won't. At the most, it will buy you access. Use it judiciously. Politicians are by and large honest and you can't buy their votes.

MAKING LAWS

Federal legislation may be requested by the President, by federal agencies, by members of Congress, by interest groups, and by you. The most astonishing thing about our government's vast and growing power is that it's fairly easy to influence—but not to change drastically. And lobbying is the most direct way. Too many Americans have the false notion that lobbying is dirty, underhanded, or even illegal. Nonsense! Lobbying is perfectly legitimate and essential if every interest in America is to be heard.

The word "lobbying" itself originates from the tradition of contacting lawmakers in the lobbies outside legislative chambers during debate to inform and influence them. Today legislators at every level look to lobbyists for accurate information on most issues. Who knows the problems of your industry better than you? Who will represent your industry's interests better than the industry itself? Industries have the *moral duty* to lobby: in a democracy **silence is regarded as acquiescence** and if you do not actively assume responsibility for making your views known, you lose the legal **right** to make them known. Failure to lobby may even in some cases lose your standing to sue in the courts, since you have failed to demonstrate any prior interest in the issue.

129

In the mid-1980s an estimated 15,000 lobbyists spend about $2 billion a year in Washington, D.C., on the gentle art of persuasion. Twenty-thousand lawyers are members of the District of Columbia Bar Association, and you can bet they're not all drafting wills and handling divorces. There are also several hundred "public interest groups" such as the Wilderness Society lobbying there too. If we are to survive, we must lobby.

Today's legislator actually depends on the lobbyist for hard facts and vigorous representation of his client's interests; not only do lobbyists know their industry's concerns better than anyone else, but also legislative staffs are never big enough to get all the background on issues by themselves. Industry trade associations have excellent lobbying talent and substantial influence, but not so much that their efforts couldn't be improved by effective citizen action to help. Activist groups have an abundance of techniques available to to influence legislation, the simplest of which is to ask their Congressmen to introduce a bill. They can write, call or visit their lawmakers, submit research findings and fact sheets, organize "convoys to Congress" for mass lobbying, and invite legislators to make speeches or debate issues.

In all these activities, it is vital for activists to demonstrate that their view has a large base of support (petitions with thousands of signatures are good evidence), has alliances with other citizen groups, and has received public attention (collections of news reports, editorials, letters to editors, and photos).

Properly made public opinion polls can also influence politicians. Concerted letter writing campaigns, however, are probably the single most vital tool of the citizen activist lobbying effort, and should be guided by their own lights, not by signals from industry's professional lobbyists. Chuck Cushman's masterful National Inholders Association lobbying program has a much better track record of wins than any industry's lobbyists because he is aggressive and a "bomb-thrower"—he will attack environmentalists head on and force them to back down or make such a controversy that legislators back down from environmentalist proposals. Cushman's "Inholder Alerts" have proven the value of letter writing campaigns. It is important to personally write by hand or typewriter every letter to a Congressman—many lawmakers just trash and don't even count form letters that citizens merely sign, although full-page newspaper ads containing hundreds of names of supportive citizens can have significant impact on key votes.

ADMINISTERING LAWS

Hassling administrative agencies has become an art with environmentalists. Federal administrative agencies control 762 million acres of land, one third of America, with the top five agencies (Bureau of Land Management, Forest Service, Department of Defense, Fish & Wildlife Service, and National Park Service in order of acreage) controlling 95%of the total. Specific management decisions are often left to these agencies by Congress, and it is crucial to keep a steady stream of political pressure on them in order to assure favorable policy. To show how important administrative decision-making can be, Dr. Carl A. Newport studied the standard day-to-day land use planning activity of the Forest Service and found that it was responsible for the loss of 20 million acres of commercial forest and additional acres of rangeland into jargony classifications like "special lands," "marginal lands," "habitat protection lands"—and as one political cartoon has it, you know you're lost when you hike into America's "Mental Health Lands"—and all of this with none of the fanfare associated with battles in Congress over much smaller wilderness area withdrawals. Watch those agencies! Fight to have federal resource lands sold to private parties!

If you want to challenge any of these bureaucratic designations and obtain administrative review, the National Forest Management Act of 1976 regulations require that you must have spoken up at the *very beginning* of the planning process when public input was first requested, and you must have mentioned the very problems you wish reviewed, or you will be denied access to the administrative review process. Nothing could stress the importance of activism more. We must pressure administrators by showing up at every meeting, every hearing, every show-me tour, and we must submit written documentation of our participation. Environmentalists do.

SETTLING DISPUTES

Courts are an important source of public policy through their power to interpret statute law. At the appeals court level, court decisions are known as "case law." The court decision itself then becomes the law of the land, regardless what the legislature originally said. Case law decisions may tone down or build up the provisions of statutory law (law made by elected legislators). In some circumstances, if a long string of cases consecutively tone a law further and further down or build it further and further up, the case law based upon these precedents may become absurdly lenient or barbarically harsh. I call this "doctrinal drift." It has happened in environmental law in only twenty years worth of cases. Prejudiced judges have taken a highly punitive stance with industry, much tougher than Congress appears to have intended.

Environmental lawyers planned it that way, and merely gave judges the arguments they were looking for. Environmental lawyers first took simple and easy cases and then, after winning more and more decisions from anti-industry or ultra-liberal judges such as those in the federal Ninth Circuit Court of Appeals, only had to defend their already conquered legal territory. Today environmental law is an overpowering force in society that strangles small business and large, creating artificial scarcities and warping free markets.

Why has American industry not achieved equally? Managers of profit-making enterprises have different goals from those of environmentalist group managers. Profit makers want to "mind the store" and assure "business as usual," those slogans that thoughtless people condemn, but that the rest of us realize as the motor of the modern world. However, this admirable decision to pay attention to creating economic goods for everyone has its drawbacks when confronted with dedicated anti-capitalist activists.

Profit makers do not want to generate lawsuits, they want to settle them. They have no program to build legal fences around the actions of environmentalist groups. They have no agenda to outlaw certain freedoms of environmentalist groups. They lobby only to protect themselves, not to destroy environmentalist power. Profit makers are entirely defensive. They have no will or ability to take the offensive to harness environmentalism once and for all. They are like boxers who will only keep up their guards but never throw a punch. Profit makers only wring their hands and complain about environmentalist excesses. They never act to regulate and destroy the power of environmentalist groups.

With such tactics, profit makers can never win. The most they can do is prevent losses. But they can never win. Because they never try. Every so-called "victory" of American industry over environmentalists consists solely of deflecting some environmentalist initiative. Not a single industry initiative to destroy environmentalist power has ever been made. Not one. Yet every political program of every environmentalist group has the result of destroying profit maker power—whether the environmentalists recognize it or not.

SUE THE BASTARDS

What to do about it? Many things. If individual firms and trade associations are reluctant to take on the environmental movement in court, citizen activists and pro-free enterprise public interest legal foundations must not be. The rise of public interest law firms that act to halt over-regulation by big government, overindulgence by courts, and excessive interference in the American way of life by extremist organizations and their attorneys is one heartening event of the past decade or so. Six regional legal centers of this nature were founded and given seed money by the National Legal Center for the Public Interest in Washington, D.C.: Capitol Legal Foundation of Washington, D.C.; Mid-Atlantic Legal Foundation of Philadelphia; Southeastern Legal Foundation of Atlanta; Mid-America Legal Foundation of Chicago; Great Plains Legal Foundation of Kansas City; and Mountain States Legal Foundation of Denver, from which Jim Watt came to the Secretariat of the Interior Department.

These organizations got their inspiration from the pioneering Pacific Legal Foundation of Sacramento, California. PLF took somewhat the same tack as the environmentalists: easy cases first (which annoyed some industrial firms who were turned down for legal aid by PLF because their cases did not turn on a public interest issue or were unwinnable) and has built up a fearsome track record of wins in striking back at environmentalism. Yet even this organization has not gone aggressively on the offensive to destroy environmentalist power in the courts. PLF takes on a wide range of cases: welfare abuse, civil liberties, consumer rights, and environmental issues, and has shown the arrogant environmentalist attorneys that the law is a double-edged sword—*Wall Street Journal* headlines have reported of these new public interest legal foundations' activities: "Environmentalists Like Developers Find Endangered Species Act Can Delay Plans," and "Businesses Are Finding Environmental Laws Can Be Useful to Them."

These pro-free enterprise legal foundations are one of our best long-term hopes for preserving our right to do business. With encouragement, new friendly public interest legal organizations will continue to appear, such as the New England Legal Foundation, the Connecticut Legal Defense Fund, the National Chamber Litigation Center (an affiliate of the U.S. Chamber of Commerce), and the Free Enterprise Legal Defense Fund (a division of the Center for the Defense of Free Enterprise, of which I am executive director).

The Free Enterprise Legal Defense Fund is at the time of this publication creating a network of experienced attorneys willing to work as co-counsel on free enterprise cases all across America. As I write, these attorneys are being trained in the tactics of environmentalists and in the aggressive pursuit of free enterprise goals: the privatization of federal lands, legal regulation of the powers of environmental organizations, and the disempowerment of government bureaucracies. Already, the case of *Center for the Defense of Free Enterprise v. U.S. Forest Service* is headed for the courts alleging violations of the Hatch Act by Forest Service officials in Nevada that would favor environmentalist takeovers of stockwatering rights. Only the future will tell of the free enterprise movement's legal effectiveness.

These pro-free enterprise legal foundations combined with

our burgeoning pro-free enterprise citizen activist groups may be the knockout punch that can defeat environmentalism. They have already worked together on defending the use of forestry herbicides, the limitation of wilderness designations, the protection of private property rights within federal areas, and many other issues. Here may be America's opportunity to push for statute and case law that will demand review of all Wilderness areas every five or ten years for possible sale to private conservation groups, that will enforce liability on environmental groups for job losses and infringement of individual rights they may cause; in short, to counterbalance the legal insistence that industry be environmentally responsible with the legal insistence that environmentalists be economically and socially responsible.

There's more to my proposed program, and it's spelled out in our final chapter.

8

The Outlook

WHERE LIES THE FUTURE of ecology wars? How we answer that question will determine whether or not American industry will be adequately prepared for the years to come. We can lull ourselves to sleep because environmentalism has slid into the background these past few years, or we can brace ourselves for action and try to recover the private property and private rights that have been seized by the federal government in the past. Many want to feel that the environmental decades are behind us, that everything has been covered in existing legislation, in clear air, clean water, wilderness, endangered species, toxic chemical, and other laws.

Powerful voices have sung this appealing song. In 1979 Congressman John D. Breaux (D-La.) said that "The 1980s may become known as the 'demand decade,' signaling new balance between industry, striving to meet public demand for needed products, and environmental factors, which have dominated legislative channels enacting regulatory controls in recent years." Breaux indeed foresaw the "supply side" emphasis of the Reagan administration. Yet he totally missed the point that existing regulations were methodically destroying basic resource industries inch by inch. He did not foresee the spreading consequences of all those environmental laws, administrative agency actions and court decisions. He could not have foretold the devastating moratorium on forestry herbicide use that would come about in 1984 based on a single obscure sentence about "worst case analysis" in EPA regulations. He could not have foretold the

137

crippling 1985 merchandise trade deficit the U.S. would rack up with Japan because of environmental laws that embargoed Alaska crude oil and federal raw log exports to Japan—which was desperate to buy them in large volumes. Breaux and all like him could not see that just putting the laws and court decisions in place would have *consequences*.

We must disenthrall ourselves from such beliefs. *Consequences* will be the dominant theme of the coming decades. Those consequences will come not only from the mechanical workings of environmental laws and court decisions, but also from the philosophy of environmentalism. They will also come from the quiet consolidation of the federal "Ecoligarchy" in the Washington power structure, even in a conservative administration. As Irving Kristol warned in *Saturday Review*, "If the EPA's conception of its mission is permitted to stand, it will be the single most powerful branch of government, having far greater direct control over our individual lives than Congress, or the executive, or state and local government." Even under the Reagan administration that is exactly what has happened. Go look at what you are obligated to do with any "hazardous" waste and ponder how the federal government arrogated such immense power to itself that it now can dictate what you do with household *cooking grease* and other hazardous wastes. Kristol was not far wrong when he said that the EPA's primary function seems to be to stop economic growth, which will inevitably result in the irony of self-imposed austerity—starvation in the midst of plenty.

THE GLOOM AND DOOM REVUE

Austerity is the favorite theme of doctrinaire environmentalists: it plays into their leadership's agenda of totalitarian control. Listen to those they read:

Arnold Toynbee, noted historian, writing in the *London Observer*: "The future austerity will be perennial and it will become progressively more severe. A new way of life—a severely regimented way of life—will have to be imposed by a ruthless authoritarian government." Of course, Toynbee was sounding a warning, but some of his readers took his statements as instructions.

Robert L. Heilbroner in *An Inquiry into the Human Prospect*: "The exigencies of the future . . . point to the conclusion that only an authoritarian, or possibly revolutionary, regime will be capable of mounting the immense task of social reorganization needed to escape catastrophe" from shortages. Again, Heilbroner was warning, but some readers were looking for directions.

William Ophuls in *Harper's*: "In a situation of ecological scarcity . . . the individualistic basis of society, the concept of inalienable right, the purely self-defined pursuit of happiness, liberty as maximum freedom of action, and laissez-faire itself all require abandonment or major modification if we wish to avoid inexorable environmental degradation and perhaps extinction as a civilization. We must thus question whether democracy as we know it can survive." It is not clear whether Ophuls was warning or advocating.

Urban planners Martin and Margy Meyerson: "A healthful and attractive environment might have to be sustained to a considerable degree by coercion . . . regulation of behavior," or "requiring individuals or firms or agencies to refrain from previous practices—practices they have come to regard from habitual usage as freedoms."

Pulitzer prize winning microbiologist Rene Dubos: "With labor, energy, and open land becoming more and more scarce and expensive, the isolated freestanding house will become an economic burden too heavy for the average person, as well as becoming socially unacceptable." Sound like Moscow?

Randall O'Toole, environmental advocate: "The next decade will see a major effort to democratize the workplace. This will be accomplished by breaking up the large timber firms into small worker-owned cooperatives through antitrust proceedings and nationalizations." Lenin couldn't have said it any clearer. In fact, Lenin endlessly complained about capitalist firms and parliamentary republics, saying they "hamper and stifle the independent political life of the *masses*, their direct participation in the *democratic* organization of the state from the bottom up," and he bragged about the Soviet Union's "millions who are creating a democracy on their own, *in their own way*" (*The Tasks of the Proletariat in Our Revolution*, V.I. Lenin, September 1917). "Democracy" in the mouth of O'Toole and his ilk appears to be one of those rubber words that means whatever he wants it to mean.

These are some of the intellectual leaders environmentalists listen to. Against their litany of authoritarian government Congressman Breaux's forecast looks naive indeed. And against the Ecoligarchy's power, the idea that industry will get a fair shake appears wildly optimistic. Remember, even Interior Secretary Jim Watt couldn't stop the federal bulldozer from stripping more private land for parks and choking the mining and offshore oil drilling industry into submission. There were too many devout environmentalists left in power through the untouchable civil service.

But the clout of environmentalism does seem to have reached a plateau in the mid-1980s. Perhaps it is vulnerable. Now is the time to strike back with citizen action to privatize the federal government, to release the iron grip of artificial regulations, to restore the free market, to starve a feeding bureaucrat. If we want a favorable outcome for American industry, we must fight for it with lobbying, litigation, pressuring administrative agencies—relentlessly. If we fight intelligently, we have a good change to win.

As we have seen in this book, beneath the healthy popular sentiment for a livable environment lies a hidden agenda of power politics among environmental leaders. Fighting these people and their program intelligently calls for a positive program of free enterprise in all arenas of politics. Now we need to complete our investigation of those arenas.

MANAGING CRISES

Crisis management is a euphemism for reacting when something bites us in the rear end (usually an unfriendly citizen activist group). Although it is a vital function and should not be scorned because it is solely defensive, it has received far too much industry emphasis while anticipatory crisis prevention measures have languished. Managing a crisis after it has erupted will always be a valuable skill because we can never predict where every crisis will come from. Managing crises usually takes the form of developing hard facts on the subjects of the crisis and getting those facts to the public. Crisis management has also traditionally involved taking legal action, such as responding to lawsuits and lobbying for a favorable outcome in legislative issues.

A more effective approach to crisis management is making "external affairs" (social and political forces) a routine part of corporate and association planning. Many purists reject this idea, feeling that business has only one legitimate social function: to meet a demand and make a profit. Milton Friedman forgive me, but that's not enough in an activist society. Economist Neil H. Jacoby, dean emeritus of the Graduate School of Management at the University of California, Los Angeles, proposes the "social environmental model" of business purposes: "Whereas both classical and managerial theory ignored the impact of political forces, the social environmental theory analyzes corporate behavior as a response to both market and non-market forces because both affect the firm's costs, revenues, and profits." In an activist society, Jacoby's model seems more appropriate and more realistic even if it smacks of heresy against the philosophy of individualism and traditional free enterprise.

THINK "SYSTEMS"

It's not heresy, just a realization that free enterprise champion Ludwig von Mises was wrong in thinking that society has no ontological existence. In striving so valiantly to defend individualism, von Mises incorrectly saw society as only the combined actions of all its component individuals. He wrote "There are no properties of society that cannot be discovered in the conduct of its members" (*Theory and History*, 1957), but this misses the point. He should have asked, "Does society have any properties that individuals do not?" The answer is, "Of course it does: institutions, governments, groups, factions, politics, and so forth." This does not repudiate the primacy of the individual as the necessary basis of society, it simply recognizes that individuals in combinations (groups) behave differently than individuals alone (the economic person): a group of individuals becomes a "system"— as in systems theory—and, as in all systems, the *interrelationships* between the *parts* (individuals) become one of the defining characteristics of the *whole* (the group).

In systems theory the whole is greater than the sum of its parts; the whole is equal to the sum of its parts *plus the interrelationships between its parts*. Think of an atom, for instance: it is not just protons, neutrons, electrons, etc., lined up in any old order such as a straight row. An atom is a system: protons, neutrons, electrons, etc., arranged in a particular relationship to each other: protons and neutrons in a central nucleus and electrons in various outer shells or probability envelopes. The specific *interrelationship* of these *parts* (numbers and arrangement of protons, neutrons, numbers and arrangement of electron shells) determine the particular properties of each *whole* atom.

The internal arrangement of an oxygen atom is only slightly different from that of a carbon atom, but that slight difference in arrangement yields enormous differences in their properties. The parts of an oxygen atom arranged in any other order would not be an oxygen atom and would have totally different properties. The *interrelationships*, the interplay of forces between the individual parts of an atom make the difference. Something analogous is true of human society. The arrangement of a society—social rules and social roles—has a decisive bearing on that society's properties.

And when those social interrelationships are those of dedicated anti-capitalist activists, the society may behave in ways highly inimical to the well-being of the individual and of free enterprise. Prof. Jacoby seems to be saying that we'd better calculate for their effects on our enterprises or we might not be around long.

THINK "EXTERNALITIES"

Many firms have made external considerations an integral part of their planning and are glad of it. Quaker Oats has a small staff that maintains personal contact with the pressure group leaders interested in its affairs, and as a result has learned to tell the real and legitimate causes from those of professional troublemakers. This helped Quaker Oats to form its policy prohibiting sponsorship of television programming containing violence or anti-social behavior.

AT&T's public relations department tries to spot problems before they become public issues and push the behavior of the corporation into line with public expectations. These and other good results indicate that other firms should carefully consider making external affairs a vice presidential position if it is not already, re-orienting public relations programs to listening as well as talking, and most importantly, making the top echelon responsive to social and political forces. Although this is still a reactive, defensive, responsive endeavor, it will vastly improve our aptitude at managing crises by taking a more anticipatory look at social and political realities. Just as some periods in history have required legal genius or financial acumen in top management, the coming years will require public affairs expertise.

DEVELOPING PUBLIC SUPPORT

This is the positive side of crisis management. It initiates action and solicits support. One of the most powerful techniques of this completely anticipatory approach is the Public Acceptance Assessment developed by Dr. Jean Mater of Corvallis, Oregon. The P.A.A. is a method of predicting what impact the public will have on a project or proposal. It includes a series of checklists that reveal what the project will contain, what impact it will have on the environment, who presently uses the site, which publics care about the impacts, how the public perceives the impacts, why certain publics care about the impacts, how much they care, whether those who care have sufficient influence to affect the outcome of the project, and which impacts can be altered. All this is done *before* the project is begun. A number of firms have used this technique under Dr. Mater's guidance and have found it highly effective.

Good works in the public interest: this is another practical approach to developing public support. Such activities as the Society of American Foresters' Urban Forestry Programs in California, and industry programs to help replant burned-over National Forest lands are prime examples. You can easily think up such projects appropriate to your own region and industry. The good works idea has built-in public relations advantages: not only can it provide a focus for positive media coverage, but it also offers opportunities to work with the more reasonable environmentalists. For example, in a replanting project, industry can offer to provide seedlings and supervisors if an agreeable environmental group will provide volunteer planters. If this proves impossible, the firm could offer a "challenge fund" to help pay for professional planters, asking for matching funds from environmentalists. The politics of such "good works" can be very touchy: the Sierra Club or the Wilderness Society cannot afford to look like they're in bed with industry, but can easily see that a "moderate, co-operative" image may do them no harm. Going to such groups asking for good works ideas will probably work better than telling them which project to volunteer for.

Wooing writers and publishers to create pro-industry literature is a largely unexplored avenue only now being looked into. Time-Life Books publishes a 24-volume series on *The American Wilderness* and *The Nature Library*, but where is *The Great Industries* series? With the majority (70 percent) of Americans now in the service sector, huge masses of people have no idea where basic wealth comes from, or how vital our resource industries and manufacturing are to their own service sector jobs. There is a crying need to get the truth about economic processes across in the same friendly, interesting word and picture essay format that Time-Life Books uses. As editor-in-chief of the Free Enterprise Press, one of my top priorities will be to solicit funds for the production of just such a mass-appeal series, among other important books.

Industry firms can also gain public support by offering show-me tours to the press and public. I've seen more minds change on the ground than in debate chambers. Also, films, slide shows, and videotapes explaining just what a firm does or plans to do have proven to be excellent public support winners. I've produced more than a hundred of them myself, and discovered that letting managers and employees of a firm tell the public what they do and how that benefits everyone prompts more general interest than you might expect. With the growing ownership of home video cassette recorders (VCRs), a free-enterprise tape library just might find its way into the burgeoning VCR tape rental market.

CITIZEN ACTIVISM

In my study of ecology wars, I have come up with two axioms: **American industry cannot save itself by itself in an activist society; and, an activist movement can only be defeated by an activist movement.**

The conclusion should be obvious: if we want to survive, we have to expand our thinking beyond economics and raise our cause to the level of a movement. As public relations counselor Philip Lesly wrote, "The power of a movement is its ability to get its cause onto the agenda of the power leaders: Congress, the administration, the government agencies, the courts. . ." American industry itself cannot become a movement: it offers nothing to join, for one thing. What we need is pro-industry, pro-free enterprise citizen activism. Citizen activism is one of the most vital parts of the spectrum of politics involved in defeating environmentalism.

Ideally, citizen activists will dovetail into existing industry programs, with industry developing hard facts and providing money, materials, and transportation, and the citizen activists becoming part of the message delivery system. Activists can also take part in highly controversial actions that industry would be well advised to stay away from, such as lobbying and litigating to require liability bonds for environmental groups that habitually delay projects and cost the economy vast sums of money. Attacking environmental groups may prove to be easier than defending industry from environmentalist attack, and it needs to be done—but by citizen activists, not by industry.

With a long and sustained effort, citizen activists may be able to pass laws that regulate environmental groups, that make them legally liable for job losses and higher prices resulting from their activities, that demand review of all wilderness areas every few years for return to multiple use status or outright privatization, and that clearly recognize under the law the legitimacy of economic concerns as being a necessary part of environmentalism. We have suffered too long with a legal system that got carried away with punishing industry for past wrongs (when they were not even recognized as wrongs). We have borne too long a legal system that fails to recognize that human civilization is a desirable part of the environment as much as the habitats of plants and animals.

SHAPING PUBLIC OPINION

The trouble with public opinion is that there really isn't any such thing. There are many publics in America today, and no one speaks for them all. American industry spends a lot of time taking public opinion surveys to see what these publics think of us. The truth is, they don't think of us much at all. When a firm gets in trouble like Union Carbide's plant in India that leaked methyl isocyanate and killed more than 2,000 people, we make headlines—as the bad guys. Otherwise, we're far in the background of daily events and anything good seldom hits the six o'clock news. So we have to go after the public if we want them to hold any opinion of us aside from the vaguely negative anti-business feelings they get by "cultural osmosis." This means taking on the hostile media (which aren't doing too well themselves in public esteem these days) in a far more enlightened way than we have done in the past.

It would not serve the nation's best interests to seek legal restrictions on the media, even though some of us who harbor a visceral hatred of them would like to see it. We must learn to beat the media at their own game. We must learn how their business operates and find out how we can fill their news needs. Instead of trying to stay out of the news, we should learn how to get into it, but on our own terms. We can get to know assignment editors of TV stations and the editorial boards of our newspapers. If we can sell them on our story, they will begin to assign reporters who actually understand free enterprise. We should be prepared with stories less than 90 seconds long, or ones that can be edited into that time slot. If we do, our message is less likely to be cut to pieces in the editing room. Short, snappy and sassy statements get the most attention and stay in people's minds longest. The best way to keep biased questions from dominating the news is to give a very long, complicated factual technical answer—editors don't like those. We need to make press conferences short. Say what we want and then leave—the longer they go on, the more likely biased questions and mistakes will appear. Our citizen activist groups can stage protests of environmentalist actions and turn the tables on our tormentors. We can no longer remain passive about the news. We must learn to make it as well as being made by it.

But news is not the only way to influence public opinion. Corporate public interest programs can have a favorable impact. A number of firms, for example Weyerhaeuser, have sponsored NBC "White Papers," an honored series of television public issue specials, and the International Paper advertising campaign in opinion leader and futurist magazines such as *Omni* dealing with the impact of social and technological innovations upon our future lives.

The feisty citizen watchdog organization Accuracy In Media has an admirable track record as a shaper of public opinion by hounding the press to get its stories straight, especially in the fields of national defense and free enterprise. Reed Irvine, AIM's redoubtable leader, is a national treasure when it comes to setting the record straight. His prodding convinced Public Broadcasting officials to show AIM's critique of the PBS documentary *Vietnam: A Television History* despite howls of protest from ultraliberals in the media. AIM's critique was narrated by Charlton Heston and drew wide acclaim from thousands of viewers, especially Vietnamese refugees who had been outraged at the PBS original that glorified the communist victory. You know Reed Irvine is effective when he garners such accolades from the press as Ben Bradlee's note: "You have revealed yourself as a miserable, carping, retromingent vigilante." Just so you don't have to go to the biggest dictionary you can find, "retromingent" means "urinating backward, as a tom cat." That has to be a World Class Insult. Of course, Reed Irvine was so proud of annoying the ultraliberal Bradlee that he framed the note and grins at it on his wall every day.

Citizen activists can be particularly helpful in shaping public opinion, through education, by creating heroes and heroines, by championing the free enterprise cause, and by giving a sense of drama to industry. Poets and sensitive writers should be encouraged to write about industry. We might be surprised by the sympathetic results. At the deepest level, we should be developing a humane pro-civilization philosophy among scholars, writers, artists, playwrights, musicians, commentators, and other bellwethers of social change.

STATESMANSHIP

Diplomacy, like war, is a continuation of politics by other means. It is a way of getting the best of an enemy without firing a shot. Diplomacy with environmental leaders is a neglected part of ecology wars. Many industry leaders have told me, "I wouldn't get in the same room with one of those son of a bitches." This could ultimately be a foolish attitude as well as an overestimation of industry's power. The environmentalists have won most of the fights. We must learn to shoot with one hand and proffer the olive branch with the other. Negotiation properly conceived might just be a sound way to thwart their overall purposes—as environmentalists well know about industry.

It is true that some environmental leaders are more rational than others, and that some situations are more susceptible of negotiation than others. Among the efforts that have produced acceptable results for free enterprise is the Joint Policy Project concept. In a J.P.P., a neutral third party takes the environmentalist and industry disputants to on-the-ground investigations of development plans. Generally the two opponents can work out a mitigation scheme acceptable to both. Then both sides agree not to take any action outside the agreement—the industry will develop according to plan and the environmental group will not obstruct the development. This has actually been done in a coal development case.

The Joint Discussion Group concept has received mixed reviews. It is a meeting format run under strict rules to prevent deterioration into a shouting match. One such industry-environmentalist Joint Discussion Group produced agreements about road building standards on Forest Service lands in the western states. As a result, industry and environmentalists lobbied together against a federal agency to lower costs and site disturbance during logging road building. However, this format appears to have reached a dead end and no subsequent agreements were reached despite much wasted effort.

Mediation services have proven to have limited success. A few industry-environmentalist disputes have been negotiated by the Mediation Institute, a Pacific Northwest mediation service operated by Dr. Gerald A. Cormick, but very few disputes meet the rigorous requirements for productive mediation: the dispute must have matured to the point of stalemate in which neither side feels it can prevail entirely with the passage of time; a trained professional mediator with credibility on all sides of the dispute is essential; disputants must pay the costs of mediation. One of Dr. Cormick's more impressive achievements is a mediated settlement between the commercial fishing industry and the offshore oil industry in Central California.

The point is that there are mechanisms available for reduced conflict. All we have to do is recognize when it is appropriate to use them, which can be tricky at best. As Dr. Jean Mater said in her book *Citizens Involved: Handle With Care!*, "Wise decisions are best made in an atmosphere of mutual respect. Polarized opinion generates more heat than light." Wisdom is an admirable goal. We should pursue statesmanship where possible. But like good diplomats, we should keep the cannon ready to hand in case wisdom and sweet reason fail us. Fighting ecology wars may be no fun, but as Homer said in *The Iliad*, sometimes "The choice is left to you, to resist or die." We may not like to defeat fellow citizens who disagree with us; hurting anybody is not a happy prospect. Great generals throughout history have known what Arthur Wellesley, Duke of Wellington, said: "Nothing except a battle lost can be half so melancholy as a battle won." When free enterprise eventually wins the ecology wars, there will be no parades or medals; all we will have to show for it is business as usual. But it is exactly business as usual that is the essence of true peace and it is business as usual that environmentalists are destroying.

L'ENVOI

I have intended this book as an arsenal of intellectual ammunition for supporters of free enterprise. My message really boils down to a few points:

Ecology wars will not go away, though they may sink out of sight like a chronic disease.

The consequences of existing environmental laws, court decisions, and regulations are more likely to get worse than to get better unless we act.

American industry has been losing.

Activism is the only way to win.

Free enterprise must become a grassroots citizen movement.

The ideas expressed here should be regarded only as an introduction to the problem. This is not a guidebook on how to solve it. That is a subject for another book. If things go right, I'll take up where this book leaves off in a future how-to-do-it volume. I think I'll call it *Defending Free Enterprise*.

But let me call you to action now. If we do nothing, we will richly deserve everything we are about to get. If we act, we may preserve values far beyond those of wise resource use, of responsible stewardship, of profitable business, of material and spiritual well-being that we free enterprisers cherish. By acting, we may combat the dark cloud of primitivism and totalitarian ambition that chokes our modern world. By acting, we may help preserve the beleaguered flame of individual liberty yet another generation and pass it on safely to a happier future world. And that is a worthy goal for any endeavor.

So the gauntlet is thrown down: Environmentalists, the future will not be your Regulation State. It will be an enlightened Free Enterprise open society. Now, America—World—let's make it happen.

Freedom lives in the hearts and minds of men and women. If it dies there, no law, no court and no Constitution can ever save it.

150

APPENDIX

Appendix

The Group of Ten

The ten largest and oldest environmental organizations have created an informal coalition of leaders they pompously call "The Group of Ten." These leaders meet periodically to plot strategies and scope problems.

1. National Wildlife Federation. Founded 1936. 4.5 million members, the largest citizen group in America. 1986 budget $46 million.

2. National Audubon Society. Founded 1905. 550,000 members. 1986 budget $24 million.

3. Sierra Club. Founded 1892. 360,000 members. 1986 budget $22 million.

4. Wilderness Society. Founded 1935. 140,000 members. 1986 budget $6.5 million.

5. Natural Resources Defense Council Inc. Founded 1970. 50,000 members. 1986 budget $6.5 million.

6. Environmental Defense Fund. Founded 1967. 50,000 members. 1986 budget $3.5 million.

7. Izaak Walton League of America. Founded 1922. 50,000 members. 1986 budget $1 million.

8. National Parks and Conservation Association. Founded 1919. 45,000 members. 1986 budget $1.7 million.

9. Friends of the Earth. Founded 1969. 29,000 members. 1986 budget $1 million.

10. Environmental Policy Institute. Founded 1972. Not a membership group. 1986 budget $1.3 million.

The Group of Ten probably has extensive overlapping memberships: one individual may belong to three or four of the ten. Thus, no accounting of the actual membership of environmental groups can be assayed. Adding all the memberships as if they were different individuals gives us a total of over 5,700,000, which is about 2 percent of the total American population in 1986 (about 238 million). However, none of the dollars is counted twice. The Group of Ten rakes in $113.5 million a year. The environmental lobby is big business.

Three other influential environmental groups that don't belong to the Gang of Ten but work closely with it are:

Environmental Action Inc. Founded 1970. 20,000 members. 1986 budget $600,000.

League of Conservation Voters. Founded 1970 as a Political Action Committee. 35,000 members. 1986 budget $1.6 million.

The Conservation Foundation. Founded 1948. No members. 1986 budget $2.9 million.

Eight Court Cases That Shaped Today's Environmental Law

1. Scenic Hudson Preservation Conference v. Federal Power Commission (December 1965). A complex case in which a group of wealthy estate owners fought power lines across their Hudson River view. The 2nd Circuit Court held that factors other than economic interest could be the basis for being an "aggrieved" person, which, with other rulings gave environmental groups legal standing to sue in defense of scenic, historical, and recreational values which might be affected by power development. Opened the gates to a flood of environmental litigation.

2. Zabel v. Tabb (July 1970). The 5th Circuit Court held that the Army Corps of Engineers had authority to deny dredge-and-fill permits not only on the basis of traditional considerations of navigation, flood control, and hydroelectric potential, but also on environmental and ecological grounds.

3. Sierra Club v. Morton (April 1972). The watershed case of Mineral King Valley in which legal rights for natural objects became an explicit theory of jurisprudence. The U.S. Supreme Court held that once a citizen or group established its direct stake in an environmental decision, the plaintiff could assert the interest of the general public as well. This case reaffirmed that being "aggrieved" is not limited to economic values, but also extends to aesthetic and recreational values as well. Every American should read the germinal essay of Christopher D. Stone, *Should Trees Have Standing?*

because the line of reasoning it spells out, which was the basis of environmental arguments in this case, represents the wave of the future.

4. Sierra Club v. Ruckelshaus (November 1972). The Circuit Court in Washington, D.C. held that the EPA had acted in violation of the Clean Air Act in approving state plans that permitted certain relaxations in standards for existing air quality.

5. The United States v. SCRAP (June 1973). The U.S. Supreme Court held that in class action suits, if the alleged harm will affect a small group of people, the plaintiff must be able to prove that he will be one of those affected; but, if the harm affects all citizens, then any citizen may bring suit. Spread standing to sue in environmental matters to more and more groups and individuals.

6. Scientists' Institute for Public Information v. Atomic Energy Commission (June 1973). The Circuit Court in Washington, D.C. held that the National Environmental Policy Act of 1969 required the preparation of an environmental impact statement even at the research stage of any federally funded project.

7. Kleppe v. Sierra Club (July 1976). The U.S. Supreme Court held that no immediate preparation of a regional environmental impact statement was required from the Department of the Interior concerning regional coal development in the northern Great Plains area since there was no federal plan or program involved. Environmentalists viewed this case as a defeat in their drive to push environmental impact statement requirements into private commercial developments.

8. E.I. Du Pont de Nemours and Company v. Train (February 1977). The U.S. Supreme Court held that the EPA has authority to establish uniform 1977 and 1983 effluent limits for catgeories of existing point sources of water pollution, provided that allowances are made for variations in industrial plants.

The Ten Most Important Books on Environmentalism

Nobody could be expected to read the two or three hundred major books by environmentalists. However, every literate American should make the effort to plow through a few key volumes in order to gain a basic understanding of the problem that now faces the nation in the form of an institutionalized force for big government and against free enterprise.

In selecting these top ten I have chosen for *importance*, not for readability. Some of this stuff is hard slogging. There's nothing for it. Just grit your teeth and read. It will illumine, it will outrage, it will enrage. I hope it will spur you to action.

Man and Nature. By George Perkins Marsh. The fountainhead of modern environmentalism, this 1864 book was the first to question the American mythos of superabundance and the inexhaustibility of the earth. As close to being *the* source of the conservation movement as any one book can be. Harvard University Press reissued this classic in 1965.

A Sand County Almanac. Aldo Leopold. 1949 Oxford University Press, Oxford. This book singlehandedly brought hundreds of thousands of readers to the environmental movement during the 1960s and '70s. Leopold's "land ethic" is fully detailed here for the attentive citizen to critically examine.

Silent Spring. Rachel Carson. 1962 Houghton Mifflin, Boston. Living proof that clever book titles can become cultural symbols, this classic of bad science and good propaganda swayed millions against forest and agricultural pesticides.

Ecotactics. John Mitchell, Editor. 1970 Simon & Schuster, New York. An inside look at the Sierra Club's pandering to the youth movement of the 1970s for environmental support. It is important for its bald-faced revelation of the bankrupt mentality of the anti-technologist.

Wilderness and the American Mind. Roderick Nash. 1967 Yale University Press, New Haven. This book provides the most useful history of our ideas and views about wilderness and how they shaped the laws and moods of America since colonial times.

Man's Responsibility for Nature: Ecological Problems and Western Traditions. John Passmore. 1974 Charles Scribner's Sons, New York. Traces environmental thought back to biblical and ancient Greek sources. It also punches holes in environmentalist ideology. Makes the reader aware of how important cultural considerations are in understanding environmentalism.

The Machine in the Garden. Leo Marx. 1964 Oxford University Press, Oxford. Clarifies the anti-technology argument. Shows how crucial literature and poetry have been in influencing legislation, an area completely overlooked by industry lobbyists.

The Forest Killers. Jack Shepard. 1975 Weybright and Talley, New York. A vicious but well-researched attack in the forest industry by a no-holds-barred investigative reporter. This is yellow journalism at its yellowest.

The Existential Pleasures of Engineering. Samuel Florman. 1976 St. Martin's Press, New York. Written by a practicing professional engineer, it is a warning to his profession that they must gain greater depth in the liberal arts and social sciences if they are to cope with the cunning attorneys and intellectuals swelling environmentalist ranks.

Environmentalists: Vanguard for a New Society. Lester W. Milbrath. 1984 State University of New York Press, Albany. The *Mein Kampf* of the environmental movement. Passionately argues for the overthrow of capitalism by the "New Environmental Paradigm." Surprisingly candid look at the real agenda of the environmental movement—the hidden agenda is no longer hidden.

BIBLIOGRAPHY

Bibliography

Abbey, Edward. *The Monkey Wrench Gang.* New York: J.B. Lippincott, 1972.

Alinski, Saul D. *Rules for Radicals: A Pragmatic Primer for Realistic Radicals.* New York: Random House, 1971.

Arnold, Ron. *At the Eye of the Storm: James Watt and the Environmentalists.* Chicago: Regnery Gateway, 1982.

Arnold, Ron. "Environmentalism, Pesticide Use, and Rights-of-Way." In *Roadside Management.* Transportation Research Record 859. Washington, D.C.: National Academy of Sciences, 1982.

Arnold, Ron. "The Politics of Environmentalism." In *Notes on Agriculture.* Guelph, Ontario: Ontario Agricultural College, University of Guelph, 1981.

Arnold, Ron. "The Environmental Movement and Industrial Responses." Paper presented at the Public Affairs and Forest Management Seminar, Toronto, Canada, 1985.

Arnold, Ron. "Ideologies and Agriculture." Paper presented at the 21st National Convention of the Agricultural and Veterinary Chemicals Association of Australia, Sydney, Australia, 1985.

Arnold, Ron, and Versnel, John. "Corporate Responsibility for Free Enterprise," in *Toledo Law Review,* Vol. 17, No. 2, 1986.

Ashby, Eric. *Reconciling Man with the Environment.* Stanford: Stanford University Press, 1978.

Barbour, Ian, ed. *Western Man and Environmental Ethics.* Reading: Addison-Wesley, 1973.

Barnett, Harold J. and Morse, Chandler. *Scarcity and Growth.* Baltimore: Johns Hopkins Press, 1963.

Bates, Marston. *The Forest and the Sea: A Look at the Economy of Nature and the Ecology of Man.* New York: Random House, 1960.

Bateson, Gregory. *Mind and Nature: A Necessary Unity.* New York: E.P. Dutton, 1979.

Bell, Daniel. *The Coming of Post-Industrial Society: A Venture in Social Forecasting.* New York: Basic Books, 1973.

Bell, Daniel. *The Cultural Contradictions of Capitalism.* New York: Basic Books, 1976.

Budyko, Mikhail Ivanovich. *Global Ecology.* (English translation) Moscow: Progress Publishers, 1980.

Callenbach, Ernest. *Ecotopia.* Berkeley: Banyan Tree Books, 1975.

Callenbach, Ernest. *Ecotopia Emerging.* New York: Bantam Books, Inc., 1981.

Carson, Rachel. *Silent Spring.* Boston: Houghton Mifflin, 1962,

Claus, George, and Bolander, Karen. *Ecological Sanity.* New York: David McKay, 1977.

Cockburn, Alexander, and Ridgeway, James. *Political Ecology: An Activist's Reader on Energy, Land, Food, Technology, Health, and the Economics and Politics of Social Change.* New York: Times Books, 1979.

Cohen, Abner. *Two-Dimensional Man: An Essay on the Anthropology of Power and Symbolism in Complex Society*. Berkeley: University of California Press, 1974.

Collingwood, R.G. *The Idea of Nature*. Oxford: Oxford University Press, 1960 (Originally published by Clarendon Press, 1945)

Commoner, Barry. *The Closing Circle*. New York: Alfred A. Knopf, 1971.

Coser, Lewis. *The Functions of Social Conflict*. New York: The Free Press, 1956.

DeBell, Garrett, ed. *The Environmental Handbook*. New York: Ballantine Books, 1970.

Dioumoulen, I. I., Ivanov, I. D., Krasnov, G. A., Polezhayev, V. N., Volkov, M. Y., and Zaitsev, N. G. *For A Restructuring of International Economic Relations*. (English translation) Moscow: Progress Publishers, 1983.

Dubos, Rene. *So Human An Animal*. New York: Charles Scribner's Sons, 1968.

Duncan, Hugh Dalziel. *Symbols in Society*. Oxford: Oxford University Press, 1968.

Ehrlich, Paul. *The Population Bomb*. New York: Ballantine Books, 1968.

Ellul, Jacques. *The Technological Society*. New York: Alfred A. Knopf, 1964.

Ellul, Jacques. *Propaganda: The Formation of Men's Attitudes*. New York: Alfred A. Knopf, 1965.

Feuer, Lewis S. *Ideology and the Ideologists*. New York: Harper & Row, 1975.

Florman, Samuel C. *The Existential Pleasures of Engineering*. New York: St. Martin's Press, 1976.

Freud, Sigmund. *Civilization and its Discontents.* (English translation by Joan Riviere.) London: Hogarth Press and Institute of Psycho-Analysis, 1930.

Friedman, Lawrence M. *A History of American Law.* New York: Simon & Schuster, 1973.

Gerlach, Luther; and Hine, Virginia H. *People, Power, Change: Movements of Social Transformation.* Indianapolis: Bobbs-Merrill, 1970.

Giedion, Sigfried. *Mechanization Takes Command.* New York: W.W. Norton, 1948.

Glacken, Clarence J. *Traces on the Rhodian Shore: Nature and Culture in Western Thought From Ancient Times to the End of the Eighteenth Century.* Berkeley: University of California Press, 1967.

Gughemetti, Joseph, and Wheeler, Eugene. *The Taking.* Palo Alto: Hidden House Publications, 1981.

Hardin, Garrett. "The Tragedy of the Commons," *Science*, December 13, 1968.

Hayes, Samuel P. *Conservation and the Gospel of Efficiency: The Progressive Conservation Movement 1890-1920.* Cambridge: Harvard University Press, 1959.

Harry, J., Gale, R., and Hendee, John. "Conservation: An Upper Middle Class Social Movement." *Journal of Leisure Research*, 1969, *1*, 246-254.

Heilbroner, Robert L. *An Inquiry into the Human Prospect, Updated and Reconsidered for the 1980s.* New York: W.W. Norton, 1980.

Hendee, John C.; Catton, Jr., William R.; Marlow, Larry D.; and Brockman, C. Frank. *Wilderness Users in the Pacific Northwest—Their Characteristics, Values, and Management Preferences.* U.S.D.A.

Forest Service Research Paper PNW-61. Portland: Pacific Northwest Forest & Range Experiment Station, 1968.

Illich, Ivan. *Toward a History of Needs.* New York: Pantheon Books, 1978.

Inglehart, Ronald. *The Silent Revolution: Changing Values and Political Styles Among Western Publics.* Princeton: Princeton University Press, 1977.

Inglehart, Ronald. "Post-Materialism in an Environment of Insecurity." *American Political Science Review*, 1981, 75, (4), 880-900.

Isard, Walter. *Ecologic-Economic Analysis for Regional Development: Some Initial Explorations with Particular Reference to Recreational Resource Use and Environmental Planning.* New York: The Free Press, 1972.

Kahn, Herman; Brown, William; and Martel, Leon. *The Next 200 Years: A Scenario for America and for the World.* New York: William Morrow, 1976.

Kortunov, Vadim. *The Battle of Ideas in the Modern World.* (English translation) Moscow: Progress Publishers, 1979.

Kristol, Irving. *Two Cheers for Capitalism.* New York: Basic Books, 1978.

Kuhn, Thomas S. *The Structure of Scientific Revolutions.* Chicago: The University of Chicago Press, 1962.

Leopold, Aldo. *A Sand County Almanac.* Oxford: Oxford University Press, 1949.

Lenin, Vladimir Illich. *Collected Works* (45 volumes). English translation of the fourth, enlarged Russian edition prepared by the Institute of Marxism-Leninism, Central Committee of the Communist

Party of the Soviet Union. Moscow: Progress Publishers, 1960.

Lovejoy, Arthur O. and Boas, George. *Primitivism and Related Ideas in Antiquity*. Baltimore: Johns Hopkins Press, 1935 reprinted 1965 by Octagon Books, a Division of Farrar, Straus and Giroux, Inc., New York.

Lovelock, James E. *Gaia: A New Look at Life on Earth*. Oxford: Oxford University Press, 1979.

Maddox, John. *The Doomsday Syndrome*. New York, McGraw-Hill, 1972.

Marcuse, Herbert. *One Dimensional Man*. Boston: Beacon Press, 1964.

Marsh, George Perkins. *Man and Nature: Or, Physical Geography as Modified by Human Action* (1864). Cambridge: The Belknap Press of Harvard University Press, 1965.

Marx, Karl. *Collected Works* (50 Volumes) New York: International Publishers, 1975.

Marx, Leo. *The Machine in the Garden*. Oxford: Oxford University Press, 1964.

Maslow, Abraham H. *Motivation and Personality*. Second Edition. New York, Harper & Row, 1970.

May, Allan. *A Voice in the Wilderness*. Chicago: Nelson-Hall, 1978.

McLuhan, Marshall. *Understanding Media: The Extensions of Man*. New York: McGraw Hill, 1964.

Meadows, Donella H.; Meadows, Dennis; Randers, Jørgen; and Behrens, William W. III. *The Limits to Growth*. New York: Universe Books, 1972.

Milbrath, Lester W. *Environmentalists: Vanguard for a New Society*. Albany: State University of New York Press, 1984.

Mumford, Lewis. *The Myth of the Machine.* In three volumes. I. *Technics and Human Development.* New York: Harcourt, Brace & World, 1967. II. *The Pentagon of Power.* New York: Harcourt Brace Jovanovich, 1970. III. *Interpretations and Forecasts: 1922-1972.* New York: Harcourt Brace Jovanovich, 1973.

Nash, Roderick. *Wilderness and the American Mind.* New Haven: Yale University Press, 1967, revised edition 1973.

Neuhaus, Richard. *In Defense of People: Ecology and the Seduction of Radicalism.* New York: Macmillan, 1971.

Novik, Ilya. *Society and Nature.* (English translation) Moscow: Progress Publishers, 1981.

Ophuls, William. *Ecology and the Politics of Scarcity: Prologue to a Political Theory of the Steady State.* San Francisco: W.H. Freeman, 1977.

Pavlenko, A. *The World Revolutionary Process.* (English translation) Moscow: Progress Publishers, 1983.

Pearce, Joseph Chilton. *The Magical Child: Rediscovering Nature's Plan for Our Children.* New York: Dutton, 1975.

Pepper, David. *The Roots of Modern Environmentalism.* London, Croom Helm, 1984.

Rivers, Patrick. *The Survivalists.* New York: Universe Books, 1975.

Rokeach, Milton. *Beliefs, Attitudes and Values.* San Francisco: Jossey-Bass, 1976.

Roszak, Theodore. *The Making of a Counterculture: Reflections of the Technocratic Society and Its Youthful Opposition.* New York: Doubleday, 1969.

Roszak, Theodore. *Person/Planet: The Creative Disintegration of Industrial Society.* New York: Anchor Press/Doubleday, 1978.

Satin, Mark. *New Age Politics: Healing Self and Society.* New York: Dell, 1979. (Contains a bibliography of 250 New Age books.)

Scheibe, Karl E. *Beliefs and Values.* New York: Holt, Rinehart and Winston, 1970.

Scherer, Donald. *Personal Values and Environmental Issues.* New York: Hart Publishing Company, 1978.

Schumacher, E.F. *Small Is Beautiful: Economics as if People Mattered.* London: Blond & Briggs Ltd., 1973.

Shepard, Jack. *The Forest Killers: The Destruction of the American Wilderness.* New York: Weybright and Talley, 1975.

Shepard, Paul, and McKinley, Daniel. *The Subversive Science: Essays Toward an Ecology of Man.* Boston: Houghton Mifflin, 1969.

Simmel, Georg. *Conflict.* (1904, 1923) In *Conflict* and *The Web of Group-Affiliations.* New York: The Free Press, 1955.

Simon, Julian. *The Ultimate Resource.* Princeton: Princeton University Press, 1981.

Steen, Harold K. *The U.S. Forest Service: A History.* Seattle: University of Washington Press, 1981.

Thomas, Lewis. *The Medusa and the Snail: More Notes of a Biology Watcher.* New York: Viking Press, 1979.

Thoreau, Henry David. *Walden, Or Life in the Woods.* Boston: Ticknor and Fields, 1854.

Tucker, William. *Progress and Privilege: America in the Age of Environmentalism.* New York: Doubleday, 1982.

Vajk, J. Peter. *Doomsday Has Been Cancelled.* Culver City: Peace Press, 1978.

Wessel, Milton R. *The Rule of Reason: A New Approach to Corporate Litigation.* Reading: Addison-Wesley, 1976.

White, Lynn, Jr. "The Historic Roots of Our Ecologic Crisis," *Science*, March 10, 1969

Witts, David A. *Theft.* La Verne, California: University of La Verne Press, 1981.

INDEX

Index